SURV 10 - Surveillance, Search and Rescue Craft

Online
13-14 January 2021

ISBN: 978-1-7138-8307-4

Printed from e-media with permission by:

Curran Associates, Inc.
57 Morehouse Lane
Red Hook, NY 12571

Some format issues inherent in the e-media version may also appear in this print version.

Printed with permission by Curran Associates, Inc. (2024)

For permission requests, please contact The Royal Institution of Naval Architects
at the address below.

The Royal Institution of Naval Architects
8-9 Northumberland Street
London, WC2N 5DA
United Kingdom

Phone: 020 7235 4622
Fax: 020 7259 5912

publications@rina.org.uk

Additional copies of this publication are available from:

Curran Associates, Inc.
57 Morehouse Lane
Red Hook, NY 12571 USA
Phone: 845-758-0400
Fax: 845-758-2633
Email: curran@proceedings.com
Web: www.proceedings.com

International Conference

SURV10
Surveillance, Search and Rescue Craft

13-14 January 2021, Online Conference

PROCEEDINGS

THE ROYAL INSTITUTION OF NAVAL ARCHITECTS
8-9 Northumberland Street, London, WC2N 5DA, Telephone: +44 (0)20 7235 4622

The Royal Institution of
Naval Architects

INTERNATIONAL CONFERENCE

SURV10
Surveillance, Search and Rescue Craft

13-14 January 2021, Online Conference

Papers published in these conference Proceedings are peer reviewed.

THE ROYAL INSTITUTION OF NAVAL ARCHITECTS
8-9 Northumberland Street
London WC2N 5DA

Telephone: 020 7235 4622
Fax: 020 7259 5912

CONTENTS

SMALL CRAFT DESIGN FOR DROP TEST: CASE STUDY 1
A Nazarov, S Tapee, P Prapset and S Chattawan, Albatross Marine Design, Thailand

IMPLEMENTING PLATFORM SAFETY REVIEWS AT THE RNLI 7
M Colegate, J Andrews, H J Phillips, M Webb and N Chaplin, Royal National Lifeboat Institution, UK

QUANTIFYING THE IMPACT THAT COXSWAIN BEHAVIOUR HAS ON WHOLE BODY VIBRATION - SIMULATOR TRIAL 17
T J Newman, MOD Naval Design Partnering Team, UK

LAB TESTING TO DETERMINE SEATS' CAPACITY TO PROTECT USERS FROM IMPACT-INDUCED INJURIES 29
PR Moxley, R Lawrence, Naval Design Partnering Team, UK
TE Coe, Naval Architect and WBV Analyst, UK

REQUIREMENTS, INNOVATION AND PROCUREMENT: A 21st CENTURY UK PERSPECTIVE 41
J Louth, Subsea Craft Ltd, UK

RNLI SEVERN CLASS LIFEBOAT LIFE EXTENSION PROGRAMME 47
N Chaplin, N Fenwick, C Smith, I Sands, T Coe, F Prini and H J Phillips, Royal National Lifeboat Institution, UK

JOINT RESEARCH PROJECT AUTOPLAN 59
N Erol, UZMAR, Turkey

WHAT NEEDED TO HYBRID? MARINE PROPULSION GREENING PRINCIPLES 67
M Zeid, BERG Propulsion AB, Sweden

FUTURE TECHNOLOGY AND PRESERVING LIFE AT SEA 71
J Dalziel, Dalhousie University, Halifax, Canada
Technical personnel with IMRF Future Technology Panel member organizations

SMALL CRAFT DESIGN FOR DROP TEST: CASE STUDY

A Nazarov, S Tapee, P Prapset and S Chattawan, Albatross Marine Design, Thailand

SUMMARY

Drop test is often used for RIBs and small craft for assessment of their structural strength, as alternative to structural calculations. The paper presents the results of applied research made for 11m patrol RIB project. Part of acceptance trials program was drop test. Hydrodynamic loads were simulated with CFD and the results were used for FEA study and optimisation of bottom structure for drop test loads. The loads resulting from drop test are compared with rulebook design loads used for structural design of small fast craft and conclusions are made about validity of such testing. The recommendations on necessary structural reinforcements for drop test are provided.

1. INTRODUCTION

Fast small craft and RIBs (rigid inflatable boats) are important part of patrol and rescue operations; they conduct missions in harsh environments and different situations involving extreme loads. Due to their small size and high relative speeds such craft are subject to unfavorable action of the elements which is often unpredictable. Thus, development of calculation techniques and testing methods for such craft is critical for their design.

From the first glance, drop test is simple and natural way to validate structural strength of the craft. However, in many cases the requirement to undergo such test can significantly override normal operational loads experienced by small craft in service. Such 'override' causes increase of craft's structural mass and loss of performance and functionality. Thus, requirements and 'design for drop test' methods should be applied carefully.

The paper presents the case study of a real project - 11m RIB where the initial requirement was to conduct drop test.

2. DROP TEST APPLICATIONS

In today's practice, drop test is used for some types of small craft:

- Small recreational craft and craft for professional use designed to ISO12215-5 [3] using alternative procedure of strength assessment.
- RIBs under ISO6185 [4] - drop tests as part of performance tests, for boats up to 8m in length;
- Free-fall lifeboats and some rescue boats which are subject to SOLAS and Life-Saving Appliances Code (LSA), IMO Resolution MSC.207(81) requirements. These craft are carried on board of large commercial vessels and are not discussed in the present paper.
- Special craft designed for 'drop' applications, such as rescue and paramilitary boats, according to client's specifications.

Under ISO12215-5 Annex D [3], drop test is used for single skin or sandwich composite, also for 'other materials' with hull length $L_H \leq 6$m, where the inner surface of the hull is accessible for inspection after the test. The ISO12215-5:2019 states that the method provides assessment 'for craft built in non-reinforced plastics, where the large deflections may not be dealt with by the theory applied' in the standard. The method assumes that safety factor is incorporated into impact loads.

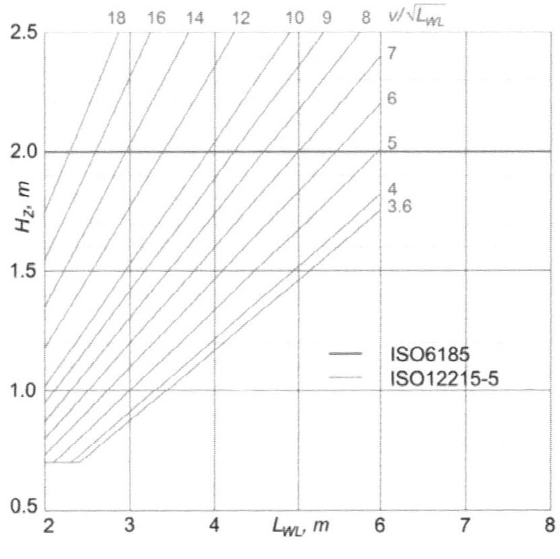

Figure 1: Summary of requirements for drop test height

Prior to the test, the craft is lifted to predefined height H_Z, which is defined by use of graph depending on speed factor $v/L_{WL}^{0.5}$, where v - maximum speed of the craft (<50kts), L_{WL} - length of waterline. Drop test heights are presented on fig.1. The boat is lifted with keel approximately parallel to the waterline, and then released. After the test, visual inspection of structure is foreseen.

In opinion of the authors of the present paper, the tests in ISO12215-5 format show potential weakness due to the fact that safety factors for composite structures and for other materials are significantly different. It has been found that drop test are primarily developed for composite craft and with present test height settings might not be suitable for, say, aluminum hulls. Secondly, visual inspection after test might not be sufficient for composite craft, and non-destructive testing should be applied.

For the RIBs, ISO6185 standards have been developed. These standards foresee that RIBs under 8m in length should undergo drop-test from the height of 2m, repeated in horizontal, bow-down 45° and stern-down 45° positions. The drop test is carried with maximum total load weight and distribution specified by the manufacturer.

3. STUDY PLATFORM

In the present paper, 11m patrol RIB was used as sample model for study. The main particulars of the craft are: length of hull/at waterline 10.6/9.6m, beam max/at chine 4.0/3.2m, mass in loaded displacement condition 5500kg, design speed 45kts, hull material - composite sandwich.

All calculations below were made for 3 bottom panels in forward, middle and aft locations. All the panels are of the same unsupported size 550×900mm - see fig.2.

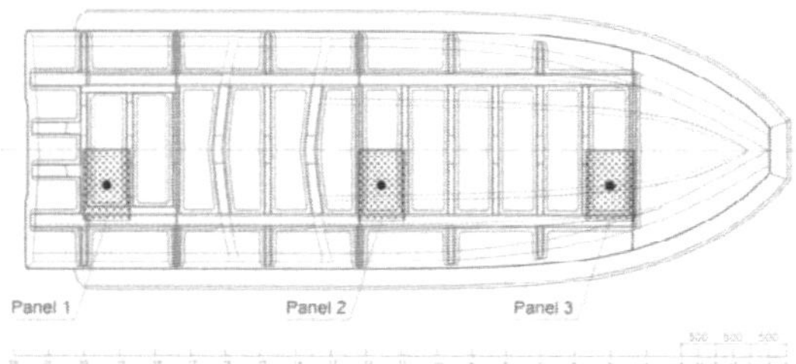

Figure 2: Diagram of bottom panels for 11m RIB

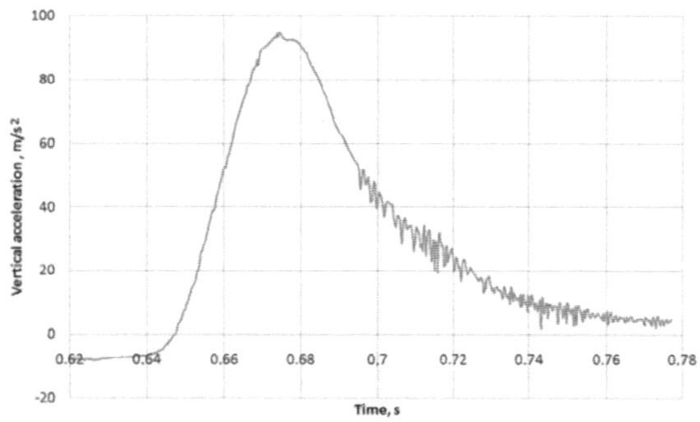

Figure 3: Drop test acceleration, 2m height

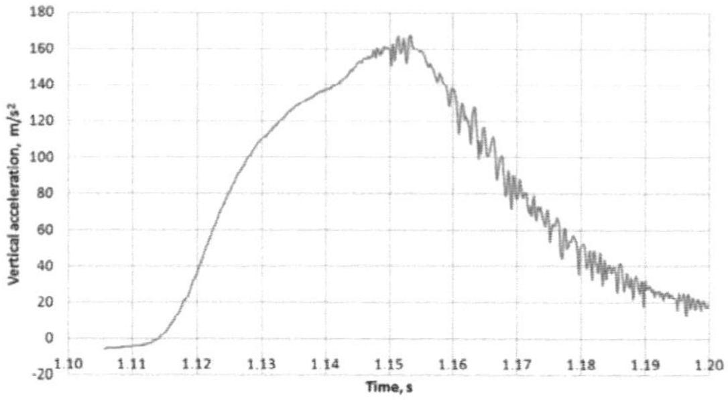

Figure 4: Drop test acceleration, 6m height

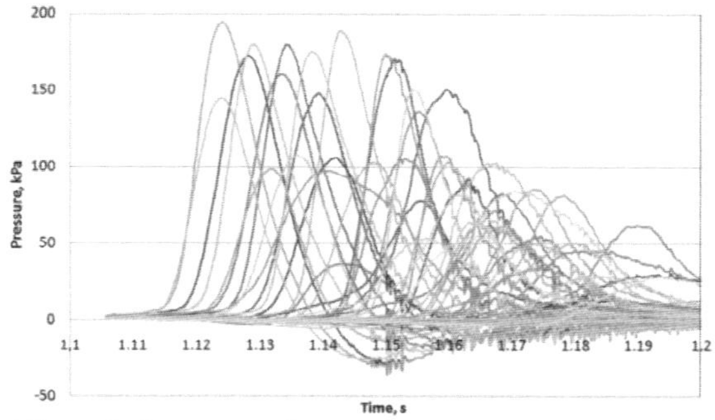

Figure 5: Summary of pressures on bottom panels, drop from 6m height

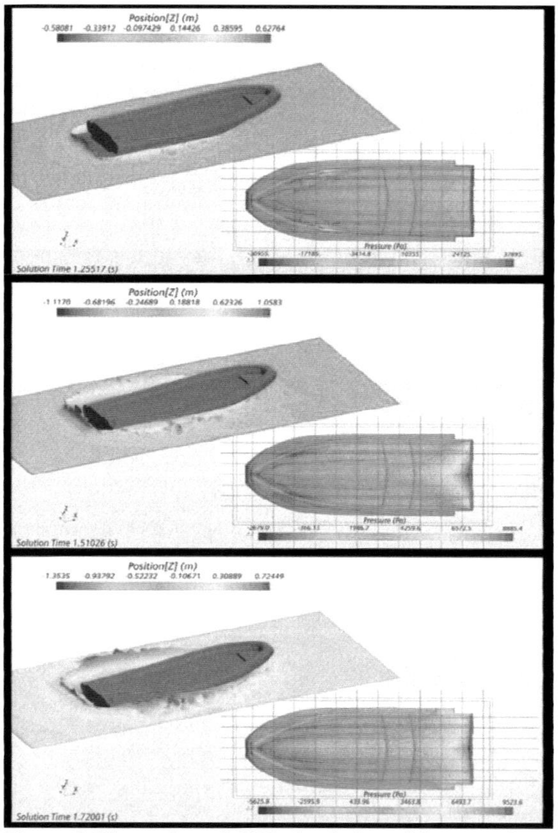

Figure 6: Illustration of CFD simulation of drop test from 6m height.

Preliminary estimates using graph from [8] indicated that drop test requirement can significantly exceed design loads defined by ISO12215-5, depending on drop height.

4. CFD SIMULATION OF DROP TEST LOADS

Drop tests were simulated with use of CFD, with variation of height 2m and 6m. For 6m drop, variation of trim angle was also studied. The acceleration records during the drop test are presented on fig.3 and fig.4. It can be seen that accelerations during drop test significantly exceed normal operational accelerations, and also can exceed typical forward collision acceleration of 12*g* [5].

Summary of pressures on panels for 6m drop is presented on fig.5. It should be noted that the most loaded area during drop test is the area around keel (centerline of the bottom).

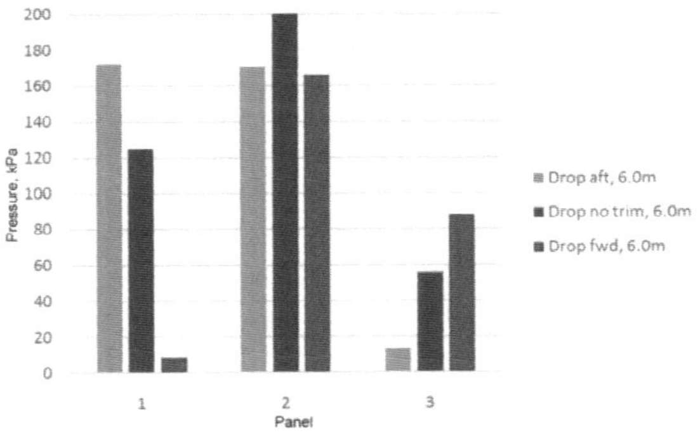

Figure 7: Effect of trim on resulting pressures, for 6m drop height.

Tests with variation of trim show significant discrepancy of pressures on studied panels (fig.7), where the case of 'aft trim' is natural free fall case. Unlike the ISO6185, in present research variation of trim did not exceed 10°. The least critical is bow down trim situation, where the primary impact is experienced by the bow part of the hull with significant deadrise.

Series of calculations was performed to compare the bottom pressures obtained during CFD simulations with 'rulebook loads', commonly used in of design fast craft (fig.8). The studied rules were:

- ISO12215-5 [3] for categories C and B;
- DNVGL 0342 Standard [2] - special standard primarily developed for commercial craft below 24m in length.
- DNVGL HSLC Rules [1]; it should be noted that by applying these rules to small fast craft the vertical acceleration is artificially limited to 6*g*.
- Lloyds Register SSC Rules [7] - found to be not workable for such fast small craft resulting unrealistically high design loads, thus not included in further analysis.

From fig.8, some useful conclusions can be made. Simulated drop test pressures for 2m and 6m significantly underestimate pressures at bow, imposed by the studied rules. However, bottom pressures for 6m height for middle and stern panels are exceeding the rulebook pressures by factor of 2..3.

It is mentioned in ISO12215-5 that 'safety factors are included in maximum impact loads'. But this seems to be the case only for composite panels on recreational craft where design stresses are $0.5\sigma_u$, $(0.50...0.65)\tau_u$ for ultimate tensile and shear stresses respectively. For metallic materials, such drop test might not be representative, as the design stresses comprise only $(0.8...0.9)\sigma_y$ of yield stress (see Table 17 of ISO12215-5 [3]).

Careful attention should be given for load comparison. Common rulebook loads are defined with 'area pressure reduction factors', which will be especially important for craft with large bottom panels.

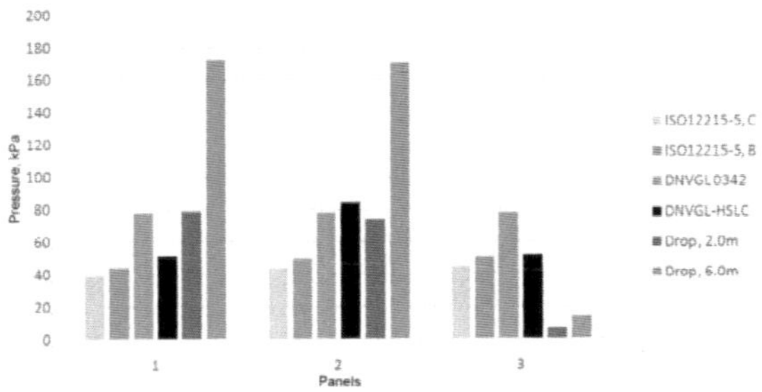

Figure 8: Comparison of rulebook pressures with CFD drop test results.

5. FEA STUDY

The results of CFD simulations of impact pressures were used for FEA study of stresses and deformations during drop test. For the craft in question, local strength is dominant during drop test on flat water. Thus, global strength study was omitted.

Some results are presented on fig.9. Reserve factor RF is used to compare the acting stresses with ultimate stresses, for Tsai-Wu stress criteria. The results of drop test from 6m is dramatic, severe damage of bottom panels and stiffeners is likely.

The results of FEA study for 2m drop case indicate that the studied craft is surviving such test. The lowest RF obtained is generally above 1.0 except some minor stress concentration points.

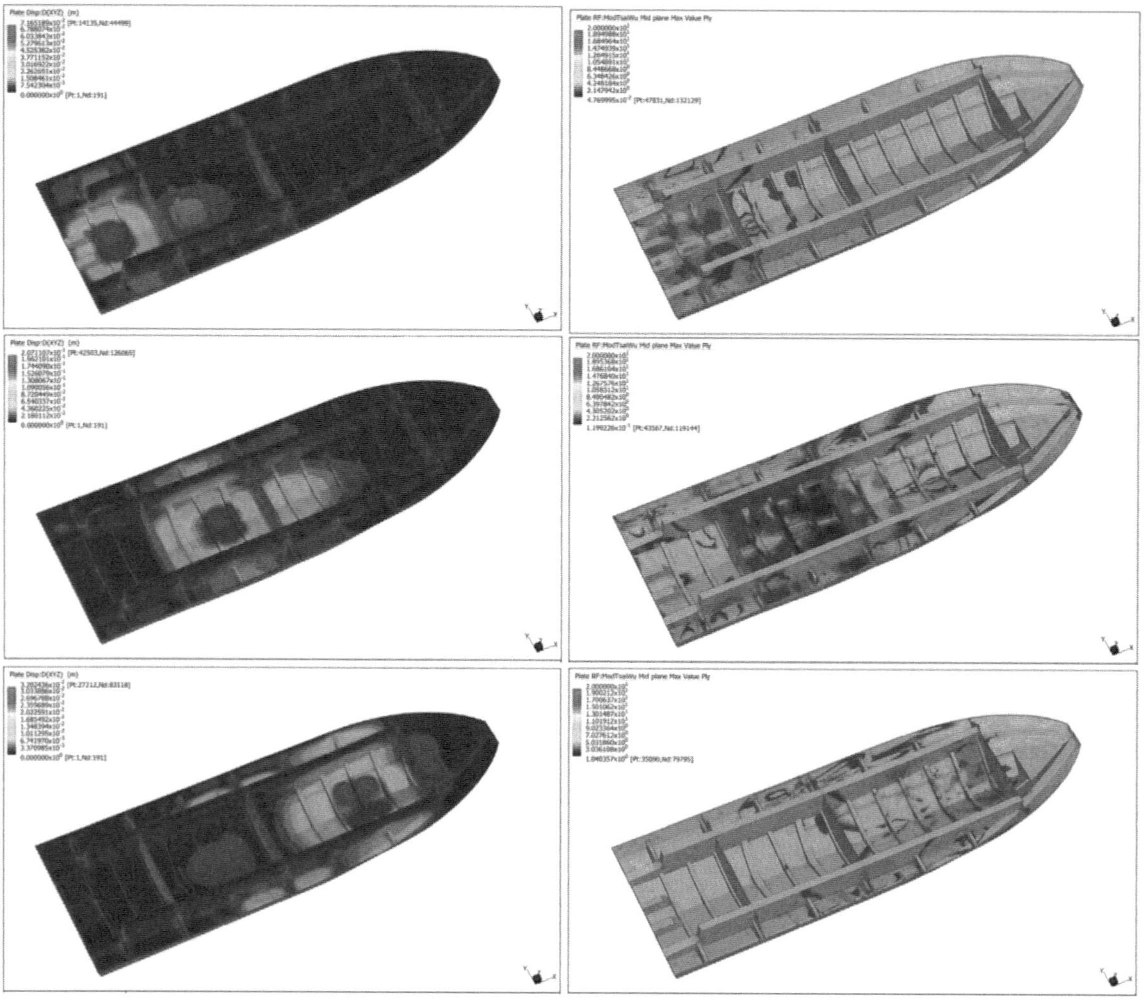

Figure 9: Results for FEA study of RIB hull for drop-test loads: left - displacements, right - reserve factors (RF) for stresses. Three consequent stages of impact are shown, free to trim drop from 6m.

6. CONCLUSIONS

Physical testing of the prototype 11m craft was performed by drop tests from 6m and 2m and the results such testing are in the agreement with the numerical modelling.

As expected, cracking local damage of the craft is highly dependent on stress concentration points. Attention to be given to structural details such as bottom steps, chines, keel/stem, transitions of stiffeners and foundations for heavy equipment.

During work on tender specifications, the operator should clearly understand the effect of drop test requirement on parameters of small fast craft. A reasonable compromise should be achieved between performance and having a heavier 'drop test proof' craft.

From the findings of the authors of the present paper, typical fast small craft in composite material designed for operational slamming loads can survive drop test from up to 2m. Exceeding this height and applying extensive drop test loads might significantly increase structural mass factors beyond the statistical limits [6].

Combination of CFD methods with FEA provides valuable information on ability of craft to pass drop test. Simplified approaches such as Wagner [8] might underestimate the pressures and resulting stresses at stress concentration points.

Further work is required from both designers and regulators to set acceptable safety factors for drop tests, for different materials.

In general, the drop test is not the best way to validate the strength of the craft designed for normal operational loads, due to differences in impact pressures, their distribution, and which have been mentioned above limitations of such tests. However, this test (and design for such test) is essential if the craft is expected to be dropped during service, for instance, while launching.

9. REFERENCES

1. DNVGL Rules for Classification: High-Speed and Light Craft (RU HSLC) - 2018.
2. DNVGL Standard 0342 - 2016.
3. ISO 12215–5:2019 Small craft - Hull construction and scantlings - Part 5: design pressures, design stresses, scantling determination.
4. ISO 6185-3:2014 Inflatable boats — Part 3: Boats with a hull length less than 8 m with a motor rating of 15 kW and greater.
5. Nazarov A. Evaluation of Acceleration Levels for Design and Operation of High Speed Craft//Morskoy Vestnik, #4-44, 2012. P.59-63. (in Russian)
6. Nazarov A. On Application of Parametric Method for Design of Planing Craft //3rd Chesapeake Power Boat Symposium, USA – June 2012.
7. Rules and Regulations for the Classification of Special Service Craft. Lloyd's Register, 2019.
8. Shenoi R.A., Wellicome J.F. Composite materials in maritime structures. Vol. 1-2. Ocean technology series, Cambridge University Press, 1993.

10. AUTHORS BIOGRAPHIES

Albert Nazarov holds the degree in naval architecture from Sevastopol National Technical University (1996) and Ph.D. in Ship Dynamics from Odessa National Maritime Academy (2004). His current position is managing director at Albatross Marine Design with experience in construction, design and research of small and high-speed craft.

Sarun Tapee holds B.Eng. degree in Naval Architecture and Marine Engineering from Chulalongkorn University (2018) At Albatross Marine Design, he is responsible for calculations of structures and stability for small and high speed craft.

Panitarn Prapset holds B.Eng. degree in Naval Architecture and Marine Engineering from Kasetsart University (2019) At Albatross Marine Design, he is involved in structural design and naval architecture of small and high speed craft.

Soravit Chattawan holds B.Eng. degree in Naval Architecture and Marine Engineering from Kasetsart University (2020) At Albatross Marine Design, he is involved in structural design of small and high speed craft.

IMPLEMENTING PLATFORM SAFETY REVIEWS AT THE RNLI

M Colegate, J Andrews, H J Phillips, M Webb and N Chaplin, Royal National Lifeboat Institution, UK

SUMMARY

With the aim to save every one, safety at sea is a fundamental goal for the RNLI. Lifesaving in an unpredictable environment is inherently hazardous, and even with the best boats, training and equipment, accidents can happen. It is, therefore, important to understand not only how safely the assets are being operated but also how that experience compares to the risks foreseen and mitigated in the asset design. The RNLI is implementing a programme of platform safety reviews (PSRs), the outcomes of which will provide the necessary assurance to accountable persons of appropriate levels of operational safety, as well as identifying opportunities for improvement. It will cover the key elements including the platform safety case, control processes, identified risks, reported incidents, status of training and competence and outcomes of assurance activities. The Shannon Launch and Recovery System (SLRS) PSR is provided as a case study.

1. INTRODUCTION (HP)

The Royal National Lifeboat Institution (RNLI) operates a lifeboat and lifeguard maritime rescue service across the UK and Ireland to save lives at sea and on inland waters. As a result of the diverse geographic and operating environment, there is a need for a large variety of different assets including boats, lifeguard equipment, buildings in addition to launch and recovery vehicles. The majority of these assets are designed, operated and maintained by the RNLI with the design authority responsibility resting within the Engineering and Supply (E&S) department.

2. SAFETY AND RISK MANAGEMENT

The RNLI's approach to the management of safety and risk within the organisation continues to evolve both in terms of overall governance and with regard to the technical assurance of assets.

The RNLI's safety team lead the co-ordination of governance activities, including incident investigations and recommendations, with the focus being on organisational learning. Each RNLI department is responsible for the appropriate safety policies, processes, procedures and associated training, as well as the management of risk. The Engineering & Supply risk registers cover the wide remit of all engineering activity, including asset related risks.

The design authority working group (DAWG), chaired by the Chief Engineer, provides the technical assurance of RNLI assets and reports into one of the RNLI's advisory committees, namely the Technical Committee comprising volunteers from a variety of industry sectors. Additional risk governance is provided by the RNLI Audit and Risk Committee and the team of internal auditors. Organisation assurance provision is described in more detail in section 4.6.

From an asset design perspective, the use of Failure Mode Effect and Criticality Assessments (FMECAs) was embedded into RNLI working processes in the early 2000s [1]. These have become increasingly important given the technology improvements made to assets and the resulting complexity this brings. Technical improvements bring increasing functionality but also risk, so the organisation has enhanced its management of this as have many industries over the last two decades. Ultimately, the safety of people, whether they be operational crews, maintainers or volunteers is the organisation's highest priority.

3. ASSET MANAGEMENT

In order to better manage the broad range of assets and equipment required to operate its lifesaving services, the RNLI is adopting asset management (AM) practices. This brings the benefits of an integrated management system approach to deliver value from assets in implementing the organisation's strategic intentions. The consideration of cost, risk and performance across the total asset lifecycle (which may extend from 5 to 100 years depending on the asset type) allows the organisation to structure long term plans for assets and the associated supporting facilities and systems.

The continual appraisal of an asset's performance (including reliability and safety) and the cost and risk levels associated with them are key activities in ensuring value to the organisation.

The RNLI has appraised itself against the requirements of ISO 55001 by undertaking an assessment against the standard asset management subjects to understand its maturity and hence develop a roadmap for improvement. Asset management objectives have been created which describe what is required to deliver Our Watch the RNLI's strategic intent, whilst considering our policy commitments. The strategies detail the work needed to achieve the objectives.

The concept of asset management is typically represented visually by way of 'The Landscape'. The RNLI interpretation of this is shown in Figure 1. Many of the key principles contained within this are discussed later in the paper.

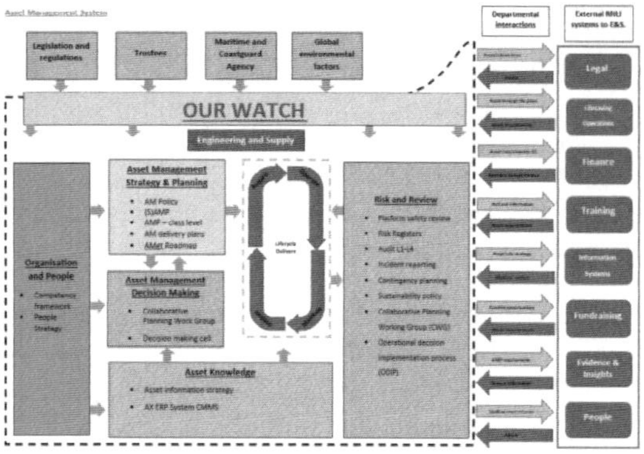

Figure 1: The RNLI Asset Management 'Landscape'

4. OVERVIEW OF PLATFORM SAFETY REVIEWS

As an organisation that provides a front-line emergency rescue service, and one that designs, builds, and maintains its own lifesaving assets, the RNLI depends upon both a paid and voluntary workforce to operate under demanding conditions to ensure asset reliability and service availability.

Conducting maritime search and rescue activities will never be entirely devoid of risk, and as such the RNLI recognises that in undertaking these voluntary activities it will face risks in order to save lives at sea. Accordingly, the relevant duty holder (or accountable person) must ensure that the organisation has suitable systems in place to be able to assess and understand the levels of risk and benefit associated with the activities and the people, assets and equipment that undertake them. Typically, the levels of risk associated with the ownership and operation of the asset will be required to be proved to be As Low As Reasonably Practicable (ALARP) [2].

This requirement to provide assessment and assurance of risk has been identified as an area for potential improvement through internal audit and associated recommendations. In seeking a methodology to provide this capability, the governing committees and their contacts were able to guide the RNLI towards the PSR process.

The PSR is a process by which a specific asset type and/or system can be assessed and reviewed for safe continued use in accordance with its designed purpose and within its intended operational environment. The PSR takes the form of a periodic audit and review of the performance of an individual asset type or system, as defined by the contents of the supporting documents used as reference material and evidence during the review.

At the time of writing this paper, the RNLI is in the early days of a PSR programme roll-out, having undertaken reviews on two operational asset types so far, and currently developing and improving the process. In due course, the reviews will form an inherent part of the institution's risk management activities and will follow a schedule of planned PSRs based on the perceived levels of risk associated with asset types (higher risks first).

The approach being adopted is to structure the PSR as an on-going performance evaluation, the focus of which is on risk in terms of original asset/system design intent, the evolution of the design intent via the engineering change processes (modifications), and operator training, use and interaction with the equipment and whether this has evolved in accordance with the changes made.

The PSR process allows the organisation to take a 'deep dive' review into the asset design and safety case documentation, and to assess the defined design intent against operational experience and recorded faults, defects and assessments to define how performance compares to expectation, and then to identify any necessary improvements to the assets, training or operational procedures in order to reduce risk levels back to acceptable levels (typically ALARP).

A general safety review of the RNLI's assets would be likely to produce a generalised, non-specific risk assessment, therefore the strategy for an individual asset PSR aims to identify the issues and risks specific to that platform – which may or may not be endemic across the asset portfolio.

The PSR is concerned with the portion of the Safety Case that the Adelard methodology describes as the safety case report [3]. It is in this report that the safety case, at the point in time of the review, is summarised, and the claims, arguments and evidence are presented to the Duty holders. Reference to supporting technical documents provides some evidence, however the analysis of the launch data, incident and defect reports provides real life information. Reviewing this data in a combined review across all departments allows us to determine if the relationship between design intent, risk assessment, operating procedures and training are sufficiently robust to mitigate our operational risks and provide adequate levels of safety.

The adoption of this methodology into the RNLI's risk management philosophy is in development and the Terms of Reference and baseline safety assessments of each asset and/or system are being created. The case study provided in the following section is, therefore, a demonstration of an early attempt to provide this baseline safety assessment and is being used to develop the approach for the future. However, the exercise has delivered good insight into the types and numbers of operational incidents and failures recorded and how they compare to what was predicted in the design stage of the equipment. The activity has also served to identify necessary improvements in both the process used to gather the pre-requisite evidence to support the claims being made, and the identification of appropriate internal/external resources needed to provide empirical evidence.

The aim of the review is to present a condensed and coherent presentation to the duty holders stating a claim of the safety level associated with the operation of the asset, supported by arguments and evidence to substantiate the claim. The process allows the duty holders (typically Engineering and Operations Directors) to assess and challenge the claims and arguments as they see fit in order to make an informed judgement on whether or not the presented case constitutes an acceptable level of risk to the safety of those working with, or in close proximity to, the asset, and that the necessary controls are in place to manage those risks.

To carry out this process effectively and in a timely manner the RNLI established a small working group of contributors comprising representation from across relevant parts of the RNLI including engineers, operators and trainers, and was supplemented with support from data analysts. It is essential that those undertaking this work have the necessary competencies and familiarity with the subject matter and risk management methods.

The activities undertaken by the working group included collating and carrying out analysis of the following:

- Safety case documentation, including operational parameters, design limitations, safety assessments and risk registers, technical operating procedures etc.
- Recorded safety incidents and investigations, defect and failure reports, asset availability and down time reports etc.
- Operational training records and information.
- Operational and technical assessments and evaluation outcomes.

Having gathered and analysed the data and information, the group developed and compiled a templated presentation to demonstrate the comparison of theoretical and evidenced risks.

4.1 SAFETY CASES

The development of asset safety cases is critical such that the RNLI can demonstrate each asset's ongoing capability to safely perform its design function, and to provide evidence that risk management has been considered in the design, and its subsequent use throughout its operational life.

Historically, consideration for design intent, (applicable to RNLI intellectual property, or commercial-off-the-shelf products), operational use, safe operation (training) and safe working practices (Standard Operating Procedures) has been an integral part of the process for new asset/product/system acceptance, or approving through life design changes.

However, a methodology for self-auditing risk management activities, to provide assurance to the accountable persons, has until recently largely been carried out as isolated activities. These are now being brought together in a more co-ordinated manner to ensure a more coherent approach is being taken.

It is noted in existing literature [3] that the safety case should take into account the needs of those who will use it to ensure safe operations and that it is essential for it to be clear and logically structured. To facilitate this, the RNLI developed a concept that captures the key elements of a safety case. This has been developed further to explain visually what the resulting PSR considerations are, as shown in Figure 2.

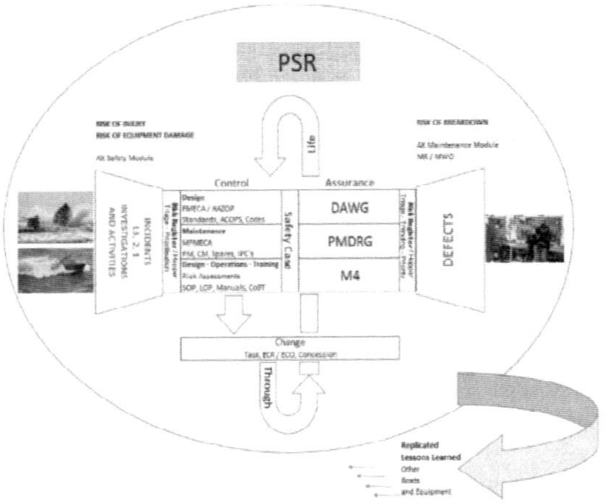

Figure 2: The RNLI's concept of PSR considerations

The RNLI has also reviewed approaches adopted by others, including Adelard [4, 5] which draws out the 'Claim, Argument, Evidence' terminologies. These can be adapted to suit the functional requirements of the organisation.

4.2 CONTROL PROCESSES

As shown in Figure 2, there are a number of control measures required to ensure the ongoing safe use of assets through their life cycle. The majority are already incorporated into standard processes, but some are in development.

The aim of the safety case approach is to ensure that the levels of risk associated with the assets is ALARP. Risks must be identified, assessed and management solutions proposed, implemented and assured. The risk mitigations fall into standard categories such as the risk hierarchy of controls (eliminate, substitute, engineering controls, administrative controls, Personal Protective Equipment (PPE)). The risk controls may also be viewed in terms of their functionality as either a design control, a maintenance control or a training/operational control.

The RNLI has created three overseeing groups aligned to these controls, and they form an essential part of the RNLI safety management system. The PSR is effectively a periodic activity which draws on the work and output of these controls to provide a snapshot of the safety associated with the asset.

4.2 (a) Design Authority Working Group (DAWG)

The DAWG comprises representation from across the asset lifecycle. The purpose of the group is to provide technical assurance of assets and to ensure that processes for design such as safety case development, engineering change, concessions etc. are defined, verified and assured and improved. This work includes the assessment of safety cases and safety control measures relating to regulation, risk, engineering standards and competences etc. The group acts as a forum for escalation of design issues requiring improvement and resolution.

4.2 (b) Planned Maintenance & Defect Reporting Group (PMDRG)

The PMDRG comprises representation from across maintenance activities in the RNLI. The purpose of the group is to oversee the safe and effective implementation of maintenance requirements, and to assure the implementation of risk controls relating to maintenance interventions.
The group reviews maintenance documentation for currency, process safety, efficiency and effectiveness. It serves as a forum for escalation of maintenance issues and to carry out analysis of defect data to propose maintenance and design improvements to improve asset reliability.

4.2 (c) Training/Operations/Engineering Group (M4)

The M4 is a tri-partite group with the purpose of ensuring that risk mitigations identified as requiring training and operational treatment are effectively managed.
It comprises representation from design authority, learning and development and operations. It provides a triple lock sign off to ensure any changes emanating from one of the represented areas is supported and approved by all parties prior to implementation.
It also serves as a forum for escalation of safety related issues from engineering, training or operations which may affect the other parties.

4.2 (d) Engineering Change

The Engineering Change process ensures that permanent modifications made to the design data of an asset comply with RNLI Design Authority (DA) policy. The DA policy safeguards against uncontrolled changes and ensures the appropriate checks (e.g. against the original design calculations and FMECA [6]) and approvals are carried out, and asset configuration and Enterprise Resource Planning (ERP) records are maintained.

4.3 RISK MANAGEMENT

The methods by which the RNLI manages risk is also an evolving area. Whilst there are standard processes within each department that allow for the identification, categorisation and analysis of risk, there are separate departmental risk registers. Whilst to some extent they perform slightly different functions, the intention is that the risk management framework will draw these closer together such that they present a more coherent current status of risk.

Specifically, the E&S asset risk register captures and categorises emerging risks that directly relate to the assets for which the Design Authority has responsibility. These risks are largely what drives the work the engineering design teams carry out to reduce the risks to an ALARP status.

4.4 INCIDENT REPORTING

Incidents and near misses across the organisation are recorded via an electronic online database. These are assigned a level of severity (1-3), triaged and actioned accordingly. The more serious incidents are classed as level 3 and result in a detailed investigation such that lessons can be learned. These may also lead to changes in operating methods, training and/or design. Any level 3 recommendations that relate to design improvements are added to the asset risk register described in section 4.3.

4.5 TRAINING AND COMPETENCE

There is a comprehensive suite of technical, standard and local operating procedures covering all RNLI assets. These are created to ensure that all equipment and systems are maintained and operated in accordance with the original design intent. Each individual person is then trained in these procedures as defined by their role and the RNLI's competency-based training framework. Their ongoing training needs are recorded and captured to ensure they remain competent in the required activities.

If any changes are required due to either an equipment change or a change in operational need, once approved, the impact on any training requirements are actioned via the M4 group as described in section 4.2.

4.6 ORGANISATION ASSURANCE ACTIVITIES

The RNLI is adopting a structured approach to assurance design and delivery consistent with the guidance contained in the standards BS EN ISO 9001:2015 (Quality Management Systems. Requirements) and BS EN ISO 19011:2018 (Auditing).

Four lines of assurance are used to focus on assessing the effectiveness of risk controls to manage risk at operational, management and organisational levels:

- Line 1 (self- check) assurance supports operational management who identify, assess, control, and mitigate risks on a day-to-day basis.
- Line 2 (management assessment) assurance assesses the effectiveness of service delivery management. These have regular management checks and reporting.
- Line 3 (process review) provides independent assurance of functional and organisational compliance risk management using internal controls and cross-discipline operational evaluation and assurance visits.
- Line 4 (audit) assurance is provided by external auditors via inspections and audits. This line of assurance reports into the RNLI's Audit & Risk Committee.

4.7 STANDARD WORK & PROCESS

Alongside the PSRs being completed on an asset by asset basis, the RNLI has also created templates and other processes to facilitate subsequent reviews. These standard documents not only provide clarity and consistency in the provision of the data required for the PSR but also provide a standard agenda for the review itself.

The standard documents and underlying processes will continue to be improved upon as further PSRs are completed.

5. CASE STUDY

5.1 BACKGROUND

The Shannon Launch & Recovery System (SLRS) comprises a tractor and carriage designed for the sole purpose of providing a mobile platform for beach launch and recovery of the Shannon class lifeboat as shown in Figure 3.

Figure 3: SLRS and Shannon Class Lifeboat. Credit: SC Innovations – Global Ltd. (All Rights Reserved)

The vehicle design encompasses two power configurations: fully powered (tractor and carriage) or tractor only (non-powered carriage). The vehicle design and prototype development were carried out over several years and with a few design partners.

Assurance of conformance with the relevant design standards was provided in the production of a Technical Construction File and a Legislation Compliance Table (LCT). An additional requirement involved a third-party safety assessment of the vehicle. This was carried out in accordance with the Joint Service Publication 454 (JSP 454) Safety and Environmental Protection for Defence. This document takes the form of a Hazard Log and focuses on human factors and design intent. As reference material for the PSR this document provided evidence of recognition of the risk present in the operational use of the vehicle and included recommendations for appropriate measures to be taken to prevent foreseeable human failures that could lead to major accidents and served as a starting point for the safety assessment to follow.

5.2 METHODOLOGY

The approach taken for the PSR was based on that described in the sections above and using the standard templates and processes already created. The data required was collected by the responsible individuals in the various teams over several weeks prior to the PSR event itself.

5.3 IMPLEMENTATION

As a newly adopted process the SLRS PSR initial Working Group proposed a structure which focused on the original design intent, followed by a study of the existing technical risks (failures in service or issues identified during production) and the manner in which these would be processed using the existing controls, to ensure the relevant changes or updates to documentation would result.

The focus then moved to assurance of risk management controls and evidencing where appropriate the efficacy of the controls, or pinpointing deficiencies in need of improvement, using operational incident data as the evidence of potential design intent / vehicle use conflict. The operational incident data then became a centre point for identifying the type, frequency and severity of the failures that had occurred as evidence of effective identification of risks and issues, the triaging of individual incidents, and the allocation of risk ownership to the appropriate department for solution provision.

However, this method failed to consider the processes being utilised to manage the incidents, and therefore whether these controls were facilitating or obfuscating the processing of the incidents to the appropriate owner and so needed adjusting to involve consideration of these controls.

The interpretation of 'evidence' for incidents rooted in behaviours was, at this stage, records of informal reminders of training, Line 1 and 2 assurance activities for the provision and use of appropriate equipment to enable safe operations, and if appropriate, refresher training courses.

For incidents rooted in technical failures, appropriate evidence of risk identification and mitigation would be provided in the form of extracts from the E&S risk register as discussed in section 4.3. Specific failure containment activities, in the form of Service Bulletins, and follow-up engineering design changes, could be evidenced and the inherent risk score in the register adjusted down to account for the mitigation activities. Where the design change activity only partially reduces the inherent risk score, further engineering work is undertaken to achieve the target score.

As the initial interpretation of adequate evidence was gathered it became apparent that a more systematic approach to presenting a coherent case was needed and that adopting the claims, argument, evidence (CAE) method would serve this purpose.

The CAE approach proved useful when evidencing the technical risk aspects of the audit, given the traceability of the engineering change process and various other E&S focused controls (Example 1 below). However, it was not as effective when applied to the task of evidencing Operational risk (Example 2).

It was identified that improvements could be made to the method used by the operators to record mitigation activities. Informal review of incidents, local triage and measures taken were not being documented in a consistent manner, thereby rendering traceability of the activity problematic.

Further review of the difficulties being encountered identified problems with how risk had, within the context of this exercise, been understood and the evidence pertaining to the risk mitigation activities identified and recorded. There was some healthy debate around this subject resulting in agreement as to what positive changes were required.

Example 1: Hydraulic hose failures

Since the vehicle entered service there have been a handful of instances in which external hydraulic hoses have failed. No injuries have been sustained, however, the risk of injury to shore crew personnel and by-standers and the environmental impact of the spills led to these incidents being give a high inherent risk score with high priority to mitigate against future occurrences.

Following the incident reporting process (Line 1 control), these incidents have been logged and triaged as a technical failure. As such the responsibility for investigation and mitigation has fallen to the Engineering and Supply department.

Within the context of Platform Safety, as an assessment of design intent and operation of the vehicle, on the surface this allocation of the risk decision is reasonable. Investigations (Level 3) into the failures have shown no operator error or violation, and there has been no evidence to suggest that human interaction has caused the failures; the hoses were fitted and secured as per design. Furthermore, there have been no injuries associated with hydraulic hose failures on the SLRS.

The investigations have instead described a pattern of failures relating to wear, corrosion and fatigue failure. Therefore, the risk profile suits the E&S department to resolve via a design intervention, and as such the risk is logged onto the E&S risk register and appropriate inherent and target risk scores are applied.

As a design issue, and if considered in isolation, the solution to the root causes might be to re-route, and/or fit anti-chafe wrapping. The inherent and target risk scores can factor in these proven solutions, with consideration paid to previous failures (likelihood), failure mode analyses including mean time before failure profiles and the target score then based on previous outcomes: hose service life extended to match manufacturer expectations.

However, within the context of platform safety, with consideration for personal safety and operational risk, and the available controls processes the risk profile changes, and the previous condition of ALARP is no longer valid.

Hose failures present the risk of serious injury resulting from high-pressure release of toxic fluid penetrating the skin or accidental ingestion. Without appropriate controls in place the likelihood of injury escalates.

The risk of harm associated with high-pressure fluid release are recognised within the RNLI and appropriate controls are in place to mitigate against this risk. Using the Swiss Cheese analogy [7], the controls processes can be considered and assessed:

- Engineering & Supply: Design intent: The vehicle is designed to perform its function of transporting the lifeboat to and from the launch and recovery site with minimal human assistance required and only one activity involving one shore crew member is required involving the hydraulic system when system is operational.

- Lifesaving Operations and Learning & Organisational Development provide specific training to the shore crew that includes standing clear of the vehicle while in use. Appropriate PPE is provided for all activities associated with the operation of the vehicle and inspections are carried out to provide assurance that this equipment is maintained to acceptable standards.

The evidence indicates that there have been no injuries or major accidents associated with hydraulic hose failures suggesting that the controls are effective. However, from the Duty Holder perspective, low to zero incidents involving accident / injury is not an indication that the risk level is ALARP.

The previous risk assessment did not account for human error, which might have led to a member of the shore crew or a bystander being near the vehicle at the point of failure. SLRS vehicle locations require the crossing of public thoroughfares to reach launch sites and, therefore, even with temporary exclusion zones during transit, the public can at times be within a few metres of the vehicle.

Experience across multiple industries has identified that it is impossible to prevent 'honest mistakes' [8]. Experience has also shown that hose failures provide little pre-warning. Corrosion witness marks and exposed inner core braid are indications of condition deterioration, but precise predictions of failure occurrence are not possible. Anti-chafe wrapping will not prevent corrosion with eventual failure stemming from fatigue failure of the outer rubber.

Therefore, the risk mitigation returns to design intervention. In this case, rather than addressing the risk in isolation and treating the root cause of the failures, the risk must consider the layered failure approach and mitigate against the severity of hydraulic oil injury and environmental contamination.

To reach the target state of ALARP, the risk was reassessed and processed through the E&S engineering change process and the design intent interrogated.

Hydraulic hose burst sleeving was applied to only a portion of the external hoses in the original design specification. This decision was based on the results of a risk assessment with the result being that only those hoses operating over a given pressure, and those requiring human interaction (quick release) were identified as requiring these additional precautionary measures. The root of this decision is the complications burst sleeving presents to maintenance and routine hose inspections. The assessment reconsidered the benefits and weighed these up against the potential for increasing the likelihood for more frequent hose failures if inspections become less effective at identifying defects.

Burst sleeving is designed to contain the high-pressure release of the oil and capture some, if not all, of the lost oil. The benefit of preventing accidental injury and the marginal gains of preventing some oil contamination weighed in favour of the change and the outcome is that all external hoses on the SLRS are now fitted with burst sleeving.

Design Authority validation of the engineering change process for the fitting of burst sleeving design requires a further review of the risk profile to determine whether the design intervention has achieved the reduction in the residual risk score to ALARP level.

With the release of the engineering change (in the form of a modification) to the regional teams and records of the modification work completed (and asset configuration records updated), the target risk score accounting for human error had been achieved and appropriate evidence was provided to the Duty Holders. However, the environmental impact of the hose failures remained at an undesirable level of risk.

Returning to the root cause analysis of the failures identified the hoses of concern as those susceptible to higher levels of bending stresses during the raising and lowering of the cradle for lifeboat launch and recovery. Having exhausted the means for protecting the hoses from contact wear and UV exposure, the remaining target area was the steel braid of the hose design. Fractures in the wire braid leading to weaknesses in the inner wall and accelerated wear from steel particle / sand abrasion were targeted for improvement. Trials involving hoses with aramid braid in place of steel braid represents work in progress to achieve the next stage of risk mitigation to ALARP.

This example demonstrates that the RNLI's controls processes are effective in identifying the layers of risk and causality that if not addressed could lead to a major incident. An improved awareness of layered risk and introducing appropriate mitigations to counter the likelihood of the 'swiss cheese holes' aligning is an area in which the RNLI is focussing improvements. The adoption of PSRs into the risk management culture is a step towards ensuring that risks are understood and considered in this context.

Example 2: SLRS Operator / Head Launcher skill fade

Repeat minor incidents involved with the operation of the SLRS vehicle have been logged on the incident database and triaged as resulting from operator and/or head launcher skill fade. This risk existed prior to the emergence of the Covid-19 virus, but an analysis of the data showed an alignment between the dates of the UK lockdown in March 2020 and an increase of these minor incidents.

The incident triage process considered the root cause to be operator and Head Launcher skill fade (human error) resulting from the curtailing of weekly station exercise activities as a result of the lockdown period. Due to perceived low risk severity levels applied during triage the individual incidents were assigned investigations levels ranging from O (no further action) to 1 (managed at regional level).

The data analysis carried out for the purposes of the review, however, led to the grouping of these incidents and the cumulative (layered) risk approach was applied from which, as demonstrated in the previous example, the risk likelihood decreases, but the severity increases. For the Duty Holder, this constitutes a risk for which appropriate mitigations are needed.

Using the same approach of assessing the risk in terms of design intent and operational use, it could be argued in this instance that design intent has no involvement, and that the failure is based on control system failures relating to training and vehicle familiarity. This is in fact how, once the numerous incidents had been grouped, the incidents were initially allocated; the burden of responsibility for providing a solution was automatically assumed to belong to the Learning and Organisational Development Team to restart station exercises (albeit at reduced frequency) and to provide assurance of crew competencies via the competency-based training (CoBT) plan.

However, design intent should always be considered in the risk assessment and triage, and the design must consider the skill level of the end-user. If the operational requirements of the vehicle are too complex, the risk of operator error exists, which when applying the cumulative risk approach could lead to bystander injury if the vehicle behaves unexpectedly. There exists therefore the obligation to design-in simplicity where possible to counteract this risk. This leads to a consideration of how best to approach these combined incidents and provide a retrospective solution: Mitigation via design to eliminate repeat minor failings may be possible, but likely to prove costly with marginal gains and, in this instance, considered to be a reaction to extra-ordinary circumstances. The alternative is to focus on behaviours and target training and vehicle familiarity.

Adopting this collaborative approach to risk mitigation enables both operational use and the design to be considered, and the option for selecting the solution that provides the maximum benefit.

6. CONCLUSIONS

The RNLI's recent experiences of developing and carrying out PSRs has so far drawn out the conclusions as follows.

The development of the PSR process is already bringing benefits to the organisation. Whilst it requires a slight cultural shift in how the RNLI manages asset risk, it presents better opportunities for more meaningful conversations based on quantified data and evidence with the results being further improvements in safety and the underlying processes.

It has been identified that the incident data could be used more effectively when combined with the asset service launch data. This information, for example the average asset hours per launch and sea conditions, can be cross referenced with those recorded on the incident database.

Departmental risk registers risks were initially reviewed in isolation. During the review, these were all reviewed to provide an aggregate view and confirm each was mitigated with all the necessary interventions in place. This approach better ensures the risks will be ALARP across the organisation and not just one area and supports the intention discussed in section 4.3 that all risk registers should eventually be connected, thus forming a coherent picture.

7. ACKNOWLEDGEMENTS

The authors wish to thank the following people who have been guiding the RNLI along the PSR journey:

- David Evans – External auditor and representative of the RNLI Audit and Risk Committee
- John Newell – Chair of the RNLI Technical Committee

8. REFERENCES

[1] BIRMINGHAM, R., SEN, P., CAIN, C., & CRIPPS, R. M., 'Development and Implementation of a Design for Safety Procedure for search and Rescue Craft', *Journal of Engineering Design, Volume 11, issue 1, Pp 55-78*, 2000.

[2] HEALTH AND SAFETY EXECUTIVE, 'Risk management: Expert guidance', Available online at: https://www.hse.gov.uk/managing/theory/index.htm (Accessed 05/01/21)

[3] OFFICE FOR NUCLEAR REGULATION, 'The Purpose, Scope and Content of Safety Cases', Office of Nuclear Regulation, NS-TAST-GD-051 Revision 7, December 2019. Available online at:

http://www.onr.org.uk/operational/tech_asst_guides/ns-tast-gd-051.pdf (Accessed 05/01/21)

[4] BISHOP, P.G., BLOOMFIELD, R.E., 'A Methodology for Safety Case Development', *Industrial Perspectives of Safety-Critical Systems: Proceedings of the Sixth Safety-critical Systems Symposium*, Birmingham, UK. Feb 1998, Felix Redmill and Tom Anderson (eds), Springer, 1998 ISBN 3-540-76189-6.

[5] BISHOP, P.G., BLOOMFIELD, R.E., 'Adelard Safety Case Development Manual' Adelard 1998. Available online at: https://www.adelard.com/papers/sss98web.pdf (Accessed 06/01/21)

[6] PHILLIPS, H. J., PITMAN, M., BOULTON, M., MÖLEN, H. E. VAN DER & STROMENGER, K., 'Design Verification Framework for Search and Rescue Craft', *SURV 9: Surveillance, Pilot & Rescue Craft, RINA, London,* April 2018.

[7] REASON, J.T., 'Human Error', Cambridge University Press (1990)

[8] SHAPPELL, S.A., WEIGMANN, D.A., 'The Human Factors Analysis and Classification System – HFACS' Available online at: DOT/FAA/AM-00/7 The Human Factors Analysis and ... (nifc.gov) (Accessed 06/01/21)

9. AUTHORS BIOGRAPHY

Michael Colegate holds the position of Compliance Engineer at the Royal National Lifeboat Institution. He is part of the Engineering & Supply team. His previous experience was Cell Leader managing the manufacture of structural composite components and the installation of mechanical, electrical and the fitting out of new build Shannon class all weather lifeboats.

Jack Andrews holds the current position of Senior Engineer (acting) at the Royal National Lifeboat Institution and is an associate member of the IMechE. He is a member of the Engineering & Supply Launch & Recovery team and specialises in the design and development of L&R assets, including the Shannon Launch & Recovery System. His previous experience includes positions as Chief and 2nd Engineer onboard both sail and motor yachts in the marine leisure industry.

Holly Phillips holds the current position of Principal Naval Architect at the Royal National Lifeboat Institution. She forms part of the RNLI's Design Authority working group and manages the naval architecture team responsible for the structural and stability aspects of the RNLI's fleet of all weather and inshore lifeboats, lifeguard equipment and other assets. Her previous experience includes that of Project Manager responsible for the design and development of the E-Class lifeboats for the River Thames.

Michael Webb Holds the current position of Principal Engineer - Launch and Recovery at the Royal National Lifeboat Institution. He is a Chartered Engineer and member of the Institute of Mechanical Engineers, responsible for the design and acceptance of technical solutions for a diverse range of equipment including marinised tractors, carriages, trailers, trolleys, davits and winches. Previous experience at the RNLI includes Contract Preparation Management and with previous employers as a Programme Manager and a Principal Engineer in Vehicle Body and Systems design and development.

Neil Chaplin holds the current position of Head of Asset Management at the Royal National Lifeboat Institution. He is a member of the senior leadership team in the Engineering & Supply department of the RNLI. He is responsible for the development and implementation of asset strategies as well as holding the role of RNLI chief engineer, responsible for technical standards of RNLI assets. His previous experience includes project manager for the RNLI Tamar class ALB design and build, Principal Naval Architect and a Staff Officer in RNLI operations.

QUANTIFYING THE IMPACT THAT COXSWAIN BEHAVIOUR HAS ON WHOLE BODY VIBRATION - SIMULATOR TRIAL

T J Newman, MOD Naval Design Partnering Team, UK

SUMMARY

A common risk to personnel is from Whole Body Vibration (WBV) and shock when transiting at speed in heavy seas, and much research has been done by maritime organisations to reduce this risk and the associated health impacts. It is well known that coxswain 'driving style' can radically affect exposure levels for a given sea state and sustained transit speed. A data-driven approach to define what makes a good coxswain from a WBV perspective is currently being developed by the Naval Design Partnering Team (NDP). In phase 1, a systematic coxswain behaviour tracking methodology has been developed and demonstrated using a motion platform-based fast craft simulator at MARIN. The performance of several experienced volunteer coxswains from MOD, RNLI and KNRM has been evaluated based on a set pattern of tests. The advantages of using the simulator, over a sea trial, have been demonstrated: it is more repeatable, more controllable, accurate and more accessible. The potential disadvantages of the approach are also discussed with reference to feedback gathered from coxswains. Analysis has shown effective throttle control is much more important than steering to reduce WBV. Several interesting trends in WBV reduction potential have been shown which it is thought, with further validation, could aid mission planning, mission execution and provide data for training autonomous feedback/control algorithms. Further work is required before the findings of this study can be fully exploited. These subsequent phases, which include sea trials, aim to provide validation and further evidence to support the initial findings.

1. INTRODUCTION

This document comprises the results of the first phase of a Naval Design Partnering team (NDP) study with the aim of improving understanding of how sustained transit speed affects shock exposure for Rigid-Hulled Inflatable Boats (RHIBs).

From a MOD procurement perspective, boat requirements are driven by defence Operational Analysis, which provides a mission profile that a boat needs to deliver. On paper, it is straightforward to set a requirement to achieve a particular transit in a given time. However, in reality there are two variables that are very hard to quantify: the sea state, and the coxswain's performance whilst navigating rough seas. Sea state is managed in the requirements by specifying a required speed in a chosen sea state, but it is difficult to predict how a boat will perform when these two variables strongly interact, and what impact that will have on shock exposure.

A large body of research has been undertaken by maritime organisations to characterise WBV/shock and associated health impacts. However, minimal research has been published on tracking coxswain behaviour in rough seas. This likely due to the technical challenges associated with human-in-the-loop trials:

- at sea – require a significant number of variables to be recorded, including the boat control signals and dynamic sea conditions, which is difficult to do accurately/repeatably, or;
- on a simulator – very few simulators exist which accurately re-create the boat physics and coxswain stimuli associated with operating in rough seas.

In terms of sea trials, Nieuwenhuis took an experimental approach to model operator behaviour on board fast RHIBs, with the aim of aiding Dutch navy procurement [1]. Three different RHIBs and operators from the Royal Netherlands Sea Rescue Institution (KNRM) and Royal Netherlands Navy (RNIN) were assessed. Using the data captured, they developed a model for throttle application and throttle duration related to the cues (e.g. wave height), to which the operator was responding. They noted the present sensor state of the art was insufficient to measure the wave height or pattern in front of the boat during the run. Instead, they inferred the wave height from boat heave and justified this using an existing boat dynamics model. They concluded from their model that the main cue for the operator to use throttle control (the most important control) is the incoming wave slope. Throttle control was applied in the wave trough and only in circumstances where the wave slope was within a certain range.

Using a simple desktop PC boat simulator, Godwin et al. studied the influence of expertise on maritime driving behaviour [2]. Eye-tracking sensors were used to measure the range of sea fixation points (i.e. field points where the eyes are looking) of expert and novice participants. They concluded that expert participants drove at higher speeds than novices, and decreased their fixation durations / increased fixation spread as wave severity increased. By connecting their findings to previous eye movement research in road driving, they suggested that novice drivers show inflexibility in adaptation to changing driving conditions. A significant limitation of this study was that the simulator consisted of a single screen with no motion platform, and therefore did not reproduce many of the coxswain stimuli that are important to study behaviour in rough seas. They also did not attempt to measure how coxswains respond in terms of the boat controls.

Neither of the above study approaches could provide an evidence base to satisfy the objectives of this study, and were therefore only taken as useful points of reference.

In order to study the ways in which a coxswain can minimise shock whilst maximising transit speed, a data-driven approach is proposed in which the above three interactions are measured or quantified simultaneously.

2. OBJECTIVES

The objectives of the overall study are:
1. Develop and demonstrate a systematic coxswain behavioural tracking methodology, with capabilities to:
a. monitor coxswain subjects control responses to a rough sea state under trial conditions;
b. monitor sea state and incoming waves which act as cues to the coxswain.
2. Undertake coxswain trials to quantify the impact sustained transit speed has on shock, with capability to:
a. analyse transit times for experienced and inexperienced coxswains in repeatable conditions for different scenarios;
b. expose coxswain to a range of realistic sea states in a controlled, controllable and recordable way;
c. determine relative importance of throttle and steering;
d. assess WBV exposure metrics and optimal behaviours;
e. relate the cues for the coxswain to trends in how the subject responds, as a function of the conditions.

The approach and trial methodology to deliver these overall objectives are outlined in Section 3 and 4, respectively. The selection of coxswains participating in the trial is outlined in Section 4.3. Section 5 reports the results and discussion. The key conclusions drawn from these results are outlined in Section 6.

3. APPROACH

In order to provide a controlled, controllable, and recordable environment where the incoming wave conditions can be selected and repeated, the application of a highly immersive simulator is considered most suitable. Various high-speed craft simulators exist and the most suitable of these has been found to be the Fast Small Ship Simulator (FSSS) developed by Maritime Research Institute Netherlands (MARIN) for the Royal Netherlands Navy.

3.1 FSSS TECHNICAL ASSESSMENT

The Royal Netherlands Navy operates 48 Fast Raiding Interception Special Forces Craft (FRISC). The FSSS was commissioned to provide a capability to exercise the boat in a simulator. This had to include the ability to accurately model hydrodynamic interactions such as broaching, surfing, planing, and capsizing. The fast craft time-domain model ('XMF/Dolphin' in Figure 1) uses a strip theory approach tuned and validated against model scale tests, sea trials and Computational Fluid Dynamic (CFD) simulations. The FSSS was built by MARIN, Cruden and Tree-C from 2015 - 2018. The simulator is located at MARIN's facility in Wageningen, Netherlands where the model scale tests were also performed [3].

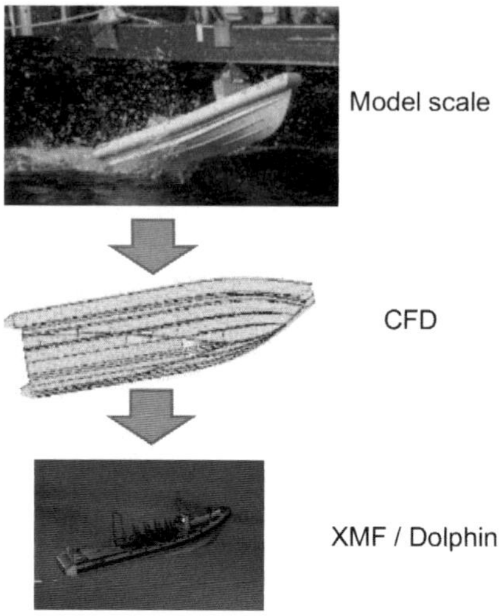

Figure: 1: Fast craft model implementation (reproduced from [3]

3.2 FSS INTERFACE DESCRIPTION

The FSSS is a highly immersive simulator which aims to reproduce the audio-visual, motion, and human-machine interface (HMI) environments of a genuine high speed craft. An overview of the FSSS, which principally comprises a 6 degree-of-freedom (6 dof) motion base, console controls and dome is shown in Figure 2.

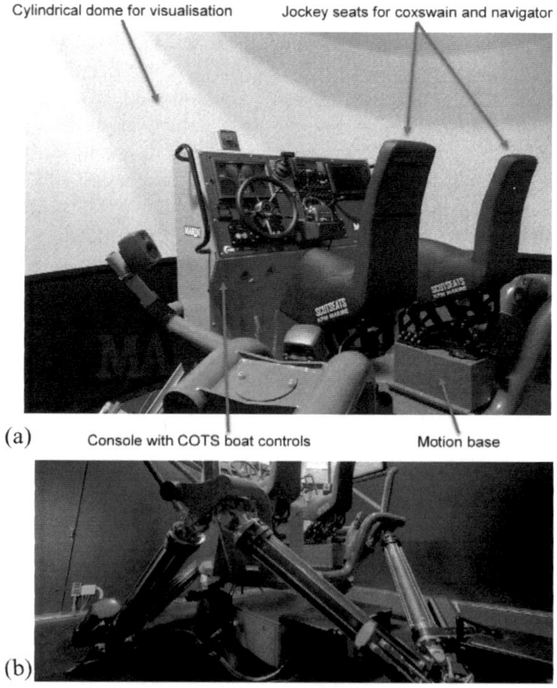

Figure 2: FSSS HMI showing (a) console controls, dome and (b) hexapod 6 dof motion base with actuators in an unextended position.

The steering, throttle, engine trim, navigation console on the motion platform are genuine boat hardware, as shown in Figure 3.

Figure 3: Console with steering wheel, throttle, engine trim and navigation screens situated on the motion base

3.3 ADAPTATIONS TO ACCOUNT FOR FSSS LIMITATIONS

A potential limitation identified in the simulator is the shortfall in peak acceleration between real impacts (of the order of 8g) and the safety limitations of the motion base (approximately 2g).

At the development stage of FSSS, feedback was gathered from experienced coxswains on whether the cues for wave slams are realistic. Visually, the cues were adjusted to provide a realistic experience to the coxswain based on their experience and feedback. In terms of peak acceleration and body forces for a heavy landing, feedback was that perhaps 'knee pain would be higher in reality'. In many respects, this is a good sign that the visual cues were correct even if the forces/pain experienced were not.

Further, the aim of this study is to monitor how a good coxswain reduces high magnitude impacts. Therefore, in theory, there shouldn't be too many impacts at the 8g level. The boat performance is still modelled at higher acceleration levels so the negative impact on transit speed and visual cues should still be realistic even if the motion base response is not.

The potential shortfall is further addressed through study design:

• Prior to the trial: briefing the coxswain to moderate their speed when reacting to rough seas compared to the way they would under real world conditions.

• During the trial: monitoring the acceleration levels being generated by the Fast Craft Model.

• Post-trial: a questionnaire to gather feedback on the coxswain's experience that includes simulator fidelity and immersion to highlight any possible caveats to the outcomes of the study.

4. METHODOLOGY

The trial set-up is outlined in the sub-sections below.

4.1 RHIB CHOICE

To minimise time and cost, and maximise the validity of the study, the existing simulator model of the FRISC was re-used. This model had the advantage that it is has undergone detailed verification and validation prior to delivery to the Dutch Navy – see the validation report [4].

The MST FRISC has a maximum speed in excess of 45 knots, and is broadly similar to the Pacific 28 employed by UK MOD. A new simulator model could have been built to represent the performance of the Pacific 28 however the additional effort required to do this was not deemed necessary as the main aim is to study coxswain behaviour, not craft performance.

4.2 GEOGRAPHICAL SETTING

The simulator was set to re-create an area of open sea with a defined sea states and test pattern. The test pattern is defined in Section 4.7 and consists of 7 way-point markers inserted at a distance of 1.5 km apart.

4.3 TEST SUBJECT SELECTION

Five coxswains were recruited for the purposes of this trial. Four experienced candidates (i.e. candidates with a significant amount of time working professionally as a RIB coxswain) were recruited from within the UK MOD Marines user groups, the Royal National Lifeboat Institute (RNLI), and the Royal Netherlands Sea Rescue Institution, KNRM. In addition, an inexperienced coxswain (someone with no experience working as coxswain, or operating a boat as a hobby) was recruited from the NDP.

In order to preserve anonymity, the following terms will be used to refer to subjects, based on their organisation of origin. The experience of each coxswain, in years, is listed below for each Ministry of Defence (MOD) or Search and Rescue (SAR) coxswain:

• MOD 1 – 12 years military experience;
• MOD 2 – 20+ years military experience;
• SAR 1 – 15 years SAR/commercial experience;
• SAR 2 – 15 years SAR/commercial experience;
• No experience – no coxswain experience.

4.4 COXSWAIN TASKING

During all runs, the coxswains were tasked to undertake transits in accordance with the test pattern defined by Figure 5 at maximum possible safe speed, given the simulated sea conditions. To attain this maximum speed and to find the optimal path through the incoming waves, the coxswains were instructed to actively control the steering and/or throttle, depending on the run type.

To exclude the influence of any initial 'learning curve' - which is required even for a highly experienced coxswain – coxswains executed a familiarisation run in advance of later runs. During the simulator familiarisation, feedback was given to coxswains on the peak acceleration of the impacts, particularly if it exceeded a reasonable limit.

4.5 COXSWAIN BRIEFINGS

The pre-trial briefing to coxswains included the following mission briefing on how to regulate their speed during the runs:

"We will not mandate a specific speed in which to complete each run, and you are expected to use your own judgement. The aim is for you to balance speed and risk according to the prevailing conditions. A good guide for this is the return to shore following a successful search and rescue operation: You should aim to get back to shore quickly to ensure the survivor can receive vital medical attention, but not so fast that you put them at additional risk."

Following the main set of runs, the coxswains were asked to complete a short de-brief interview to gauge their impression of the quality of the simulator and the shock mitigation techniques employed. Coxswains were then given the opportunity to experiment with the simulator - Figure 4.

Figure 4: Post-trial experimentation with one of the experienced coxswains

4.6 CONTROL CONDITIONS

Given the aim is to investigate the influence of both control types in isolation by excluding one control type at a time, three control permutations were tested:

• Free/free – both control types (throttle & steering) are allowed.
• Throttle only - The coxswain is tasked to use the steering wheel as little as possible. Only yaw corrections are allowed to change course once each marker has been reached.
• Steering only - At the start of the run, the coxswain is tasked to set the throttle in the right position so that the RHIB travels with maximum acceptable speed. During the run they are only allowed to change this throttle position while transitioning from one heading to another. Steering control is allowed.

4.7 TEST PATTERN

The test pattern ensured coxswains travelled in every cardinal approach to the prevailing wave direction, i.e., following seas, quartering seas, etc. and is shown in Figure 5. Markers were placed in a pattern at a set distance of 1.5 km apart such that each 'leg' (i.e. moving from one marker to the next) is the same distance. Coxswains were required to go around the outside of the markers as indicated.

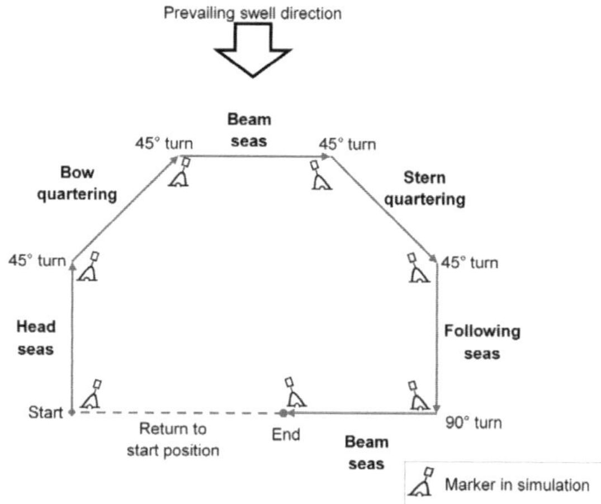

Figure 5: Test pattern

4.8 SEA STATES

An undisturbed, spectral-based deep-water wave model was used with separate wave systems for 'sea' (wind waves generated locally) and 'swell' (longer wavelength waves generated far away). The wave spectra used to generate each sea state was a Joint North Sea Wave Project (JONSWAP) spectrum set according to definitions given in Table 1.

Sea state	Significant wave height (m)	Wave period (s)	Wind speed (kts)
2	0.3	4.5	8
3	1.0	6.0	12
4	2.0	8.0	15

Table 1: Sea state (prevailing sea) definitions

Superimposed on the sea spectrum, there was constant swell of 0.5 m significant wave height and 12s period at a 10-degree offset from the prevailing sea direction.

4.9 RECORDED DATA

A range of quantities in the simulated environment and on the motion base were recorded. These are listed in Table 2. During the trial, raw time series logs of these quantities were written to a file.

Quantity	Location(s)	Data
Incoming wave height	Moving points located 1 m, 2 m and 3 m ahead of the craft bow.	Sea surface elevation
Incoming wave direction	Same as above	Sea surface wave direction of travel relative to the craft
Acceleration	Coxswain position	Accelerations in x, y, z-directions
Craft speed	Craft CoG (Centre of Gravity)	Velocity in x, y, z-directions
Craft heading	Craft CoG	Heading
GPS position	Craft CoG	Geographical position trace of craft
Throttle position	Throttle	Percentage application of the throttle.
Steering position	Steering wheel	Angle of the steering.
Boat force contributions	Body force contributions	Forces in x, y, z-directions as output by the boat model
Acceleration (real-world)	Coxswain position on the motion base	Accelerations in x, y, z-directions
Video recording (real-world)	Coxswain position	'Fly-on-wall' video

Table 2: Data recorded during the trials.

4.10 TRIAL SCHEDULE

NDP obtained a period of 2.5 simulator days. This excluded time to initialise the simulator and install real-world sensors, but included time for briefings, before and after the runs with trial participants. Each participant was given a half-day slot. The trial took place between 3rd – 5th March 2020.

4.11 POST-PROCESSING

Each coxswain undertook nine standard runs corresponding to three sea states and three control conditions. Several additional runs were also executed although these are not reported here for brevity. Each run was output to a separate run logging file. This Section provides a high-level description of processing of raw time histories of the quantities described in Table 2. All processing was automated using MATLAB routines.

4.11 (a) Spatial windowing

One of the first steps in producing the results was to remove the following unwanted parts of the time histories:

• Corners: where the heading changes significantly from e.g. head seas to bow quartering seas,
• Initial acceleration and 'false starts': the boat getting up to speed at the run start and occasional short delays in starting a run where the boat could drift,
• Overshoots: at the end of the run, some coxswains stopped the boat while others carried on.

As illustrated in Figure 6, these unwanted time history Sections were identified based on spatial windows shown in red. This resulted in six separate legs for the different headings of approximately equal distance covered.

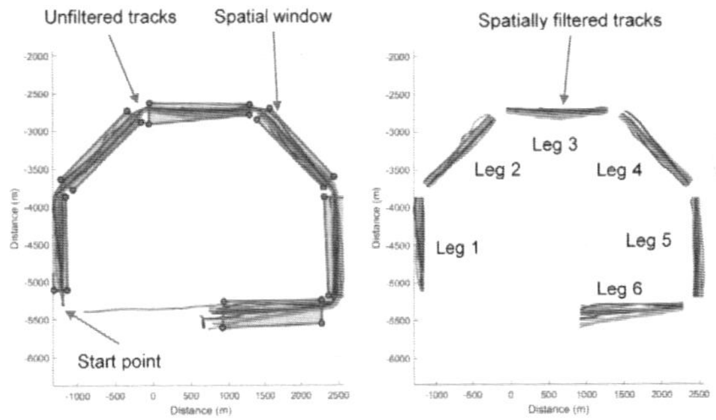

Figure 6: Splitting of unfiltered tracks using spatial windows to remove corners & initial acceleration at run start

4.11 (b) Vibration Dose Value (VDV)

The most common ways to describe the vibration experienced by a human occupant are the root-mean-square (rms) acceleration, the 8-hr normalised level (A(8)), and the Vibration Dose Value (VDV). For many vibration applications, the use of the A(8) value is preferred, as it provides a reference value which can be used to easily demonstrate the way exposure increases as a function of time relative to action and limit values.

The A(8) value, however, is not well suited to evaluating the exposure to combined shock and vibration, as it is insensitive to one-off, high-magnitude events typically associated with high speed craft operation. Acceleration time histories containing shock events are better represented by the Vibration Dose Value, which is an effective dose-response relationship between the human body and the input acceleration.

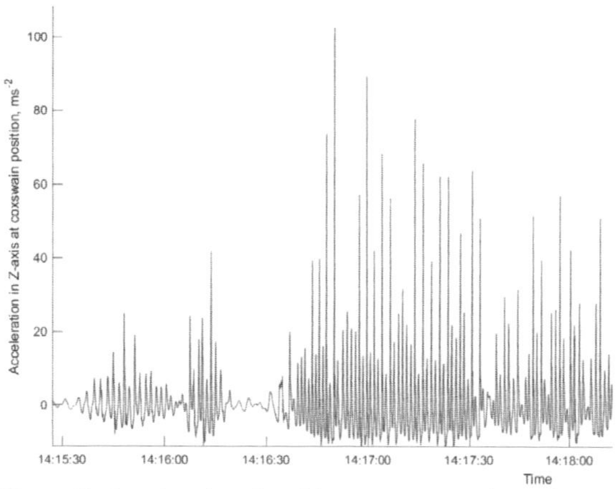

Figure 7: Acceleration time history at coxswain position

An example of the vertical acceleration trace from the simulator is shown in Figure 7.

The VDV of the acceleration time history is calculated as the integral of the fourth-root of the weighted acceleration time history raised to the fourth power. This is shown in equation format below:

$$VDV = \left(\int A(t)_w^4 \, dt \right)^{\frac{1}{4}}$$

where $A(t)_w$ is the weighted acceleration time history, with the exact weighting depending on; direction of stimulus, point of input to the human body, and purpose of the vibration assessment, and; the integration is made with respect to time. For the purposed of this study, as we are considering vertical vibration up into the human body only, the W_k weighting factor is applied, as defined in ISO 2631-1.

4.11 (c) Time to Limit Value (TTLV)

The Time To Exposure Limit Value, TTELV or TTLV, is back-calculated from the overall VDV for the given period to generate the amount of time the activity can be carried out for before an occupant is above the exposure limit value. For the purposes of this study, the following equation is used based on a daily limit value of $VDV_{threshold}$ = 21 ms$^{-1.75}$:

$$T_{VDV} = T \left(\frac{VDV_{threshold}}{VDV} \right)^{\frac{1}{4}}$$

where T_{VDV} is the time taken to reach the $VDV_{threshold}$ and T is the duration over which the VDV has been calculated.

4.11 (d) Impact-by-impact analysis

An algorithm was developed to identify and extract the conditions at each impact. Impacts are identified by being acceleration maxima over 2 m/s^2. Sections of the incoming sea surface elevation time history are cross-correlated against the acceleration time history to establish causality between acceleration peaks and a specific wave period as shown in Figure 8. The wave height is defined as trough-crest upward-crossing distance. The wave slope is the instantaneous steepness of the sea surface elevation.

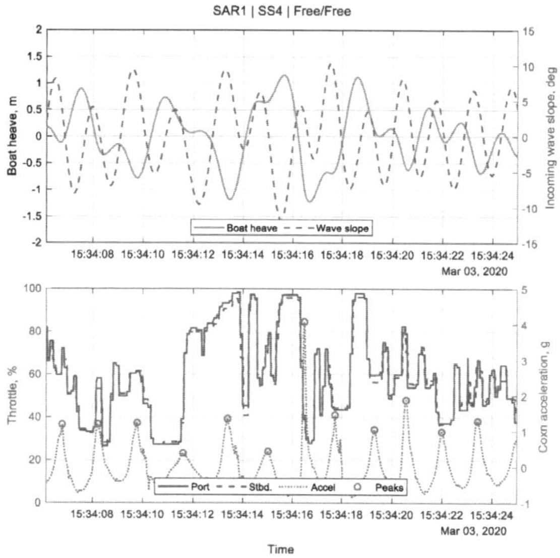

Figure 8: Impact-by-impact time series analysis

5. RESULTS & DISCUSSION

5.1 OVERALL PERFORMANCE

The overall VDV, transit time and distance travelled for each coxswain was calculated along with the sustained transit speed and Time to Limit Value (TTLV). Figure 9 shows that generally a higher sea state leads to higher VDV as expected. This is due to larger amplitude and/or more frequent shock events. The wave height increase means the boat drops from a greater height when leaving the crest of one wave, and hits the next wave resulting in a greater force on the boat and therefore acceleration at the coxswain position.

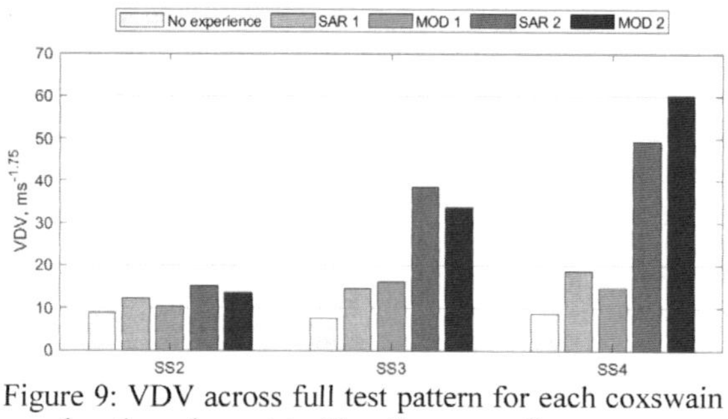

Figure 9: VDV across full test pattern for each coxswain as a function of sea state (Free/free control)

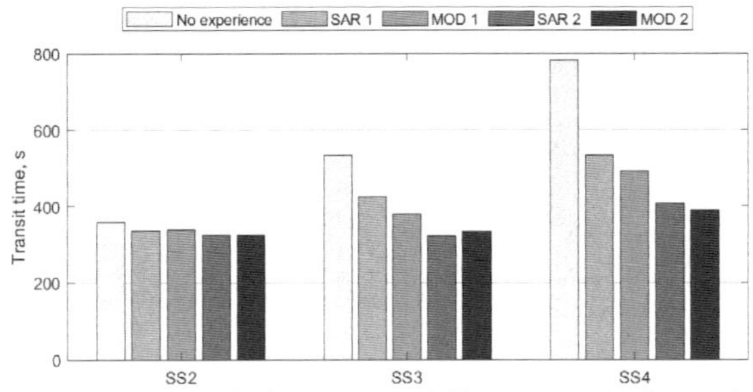

Figure 10: Transit time across full test pattern for each coxswain as a function of sea state (Free/free control)

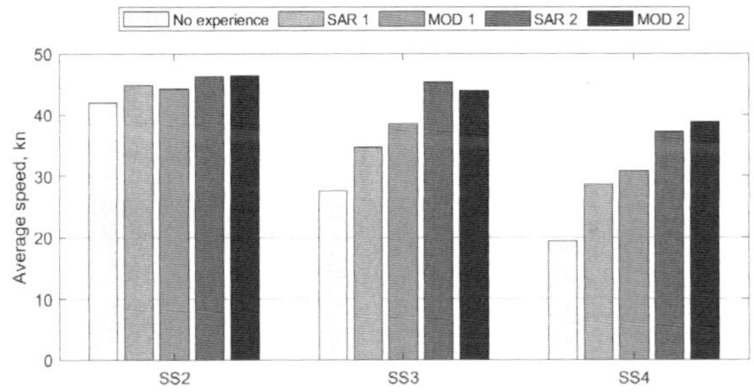

Figure 11: Sustained transit speed across full test pattern for each coxswain as a function of sea state (Free/free control)

It is important to note that, for all the results in this sub-section, the performance is taken over the test pattern as a whole. The test pattern takes 5-10 minutes to complete (excluding corners) and includes six different headings over an equal distance. As discussed further in Section 5.2, certain headings (e.g. head or bow quartering seas), have a dominant contribution to the VDV and will have been executed at a lower speed than other headings.

For a given sea state, there is a large variation in the VDV for different coxswains, even excluding the inexperienced coxswain. This is due to differences in the way each coxswain chose to control the craft. Some coxswains (SAR2 & MOD2 in particular) chose a significantly more aggressive speed (Figure 11), which led to lower transit time (Figure 10) at the expense of higher VDV (Figure 9). Clearly, although all coxswains were given the same brief, the interpretation of that brief by the experienced coxswains varied. It is postulated that, although all coxswains were all trying to balance risk in terms of achieving the mission quickly while exposing the 'casualty' to minimal unnecessary risk from shock, the experience and perception of risk was different for the SAR2/MOD2 coxswains compared to the SAR1/MOD1 coxswains. This led to a significant difference in sustained transit speed, and therefore VDV and transit time. The most risk averse and least confident was the inexperienced coxswain who had very long transit time and low VDV.

Plotting the range of transit speeds against the corresponding VDVs for each sea state, Figure 12 demonstrates that, in general, higher speed leads to a higher VDV. There are a few minor exceptions, such for SS4, where there is a slight decrease in VDV at a higher transit speed between the SAR1 and MOD1 coxswains. In this case, the MOD1 coxswain's control was clearly 'more optimal' than the SAR1 coxswain's. More effective control of the boat in terms of application of the throttle/steering is thought to be behind this, and this will be analysed later. Overall, speed is the biggest factor in determining the trend in VDV, and the selection of an 'optimal' speed will also require careful consideration of the risks of a mission as a whole (e.g. other factors that could lead to higher risk for the crew if the mission was accomplished more slowly). Further analysis to aid this decision process is given below.

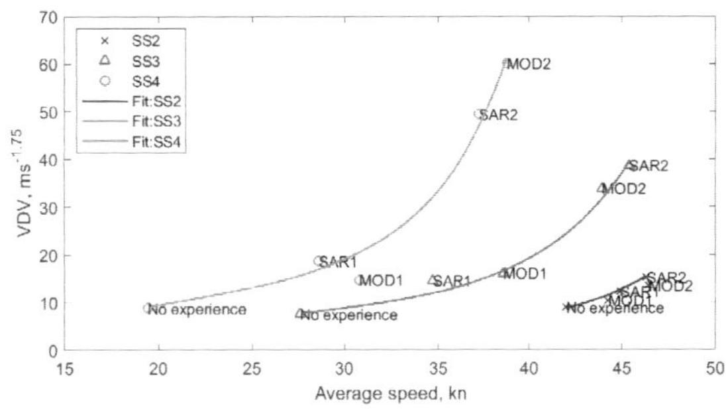

Figure 12: VDV as a function of sustained transit speed across test pattern for each sea state (Free/free control)

The data of Figure 12 can be normalised to highlight the benefit of reducing speed on minimising VDV and therefore shock exposure. Figure 13 plots the percentage reduction in VDV as a function of percentage reduction in speed. The percentage reductions are relative to the 'worst-case' coxswain with the highest VDV / speed for each sea state. There is a strong general trend (dashed line) in decreasing VDV with decreasing speed that applies independently of sea state. An equation of best fit is included. This shows the following:

• Initially a very steep curve: a small decrease in transit speed leads to a large decrease in VDV → a 10% speed reduction gives a 40% reduction in VDV.
• Diminishing returns: beyond a certain point, reducing the speed as the mission will take much longer but there is little to be gained in terms of VDV reduction.

Certain coxswains outperformed or underperformed by 5-10% compared with the general trend outlined above. This is attributable to the behaviour of individual coxswains in the way that they controlled the craft to minimise WBV whilst maximising transit speed, or otherwise.

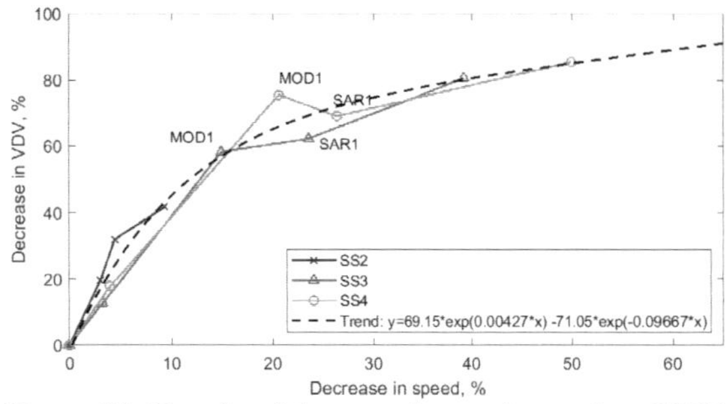

Figure 13: Benefit of decrease in transit speed on VDV relative to 'worst-case' coxswain (Free/free control)

The TLLV times vary greatly for each coxswain and sea state. In rough seas, the limit value is reached with only a few heavy (~8g) wave impacts and this is reflected in the results of the trial, particularly for the SAR2/MOD2 coxswains. Figure 14 shows the trends in TTLV as a function of speed for each sea state:

• Even in SS2, it shows the limit value is exceeded in less than an hour of continuous operation at a range of headings at the maximum speed.
• In SS3, to avoid exceeding the limit value in an 8 hour day, the speed would have to be reduced from the maximum ~45 kts to around 30 kts.
• In SS4, even at 30 kts instead of the maximum observed speed of ~40 kts, the limit value is exceeded in approximately 30 minutes.

The trends in seen in Figure 14 are approximate but compare favourably with those presented in the WBV DIN [5] for MOD high-speed craft. They could be further validated and extended by gathering more data using the simulator with different coxswains operating craft at different speeds and sea states. It should also be noted that the overall performance shown here represents a 'cumulative average' over the different headings. The TTLV could be even lower for continuous operation in, for example, head seas and this is discussed further in Section 8.2.

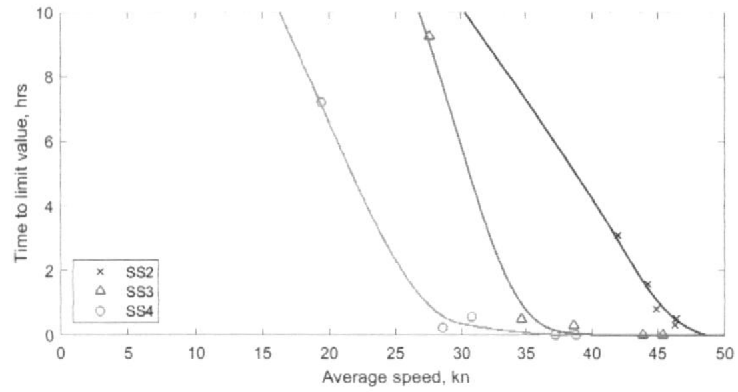

Figure 14: Time to limit value (TTLV) as a function of speed for each sea state

5.2 OVERALL VARIATION WITH HEADING

The approximately 8 km test pattern is made up of six headings, starting with head seas and finishing with beam seas. The results in this Section show how the same key metrics from Section 8.1 broken-down for each heading.

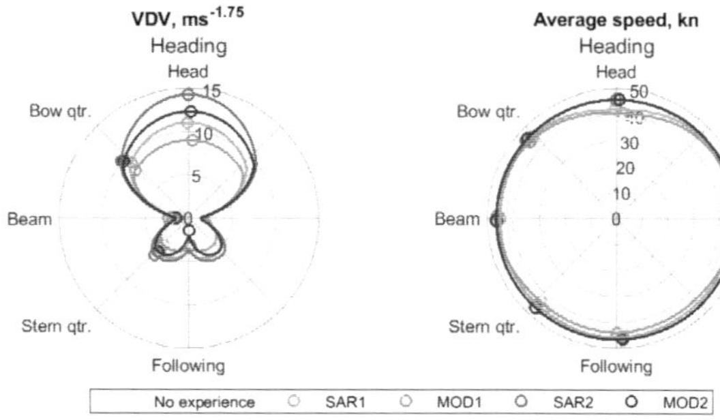

Figure 15: VDV / Transit speed variation with heading for each coxswain in Sea State 2 (Free/free control)

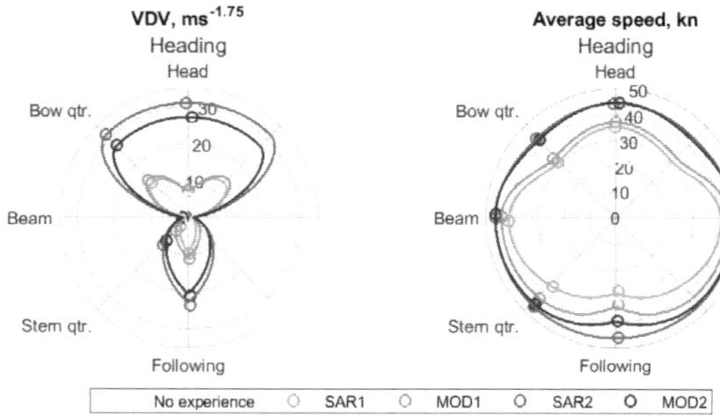

Figure 16: VDV / Transit Speed variation with heading for each coxswain in Sea State 3 (Free/free control)

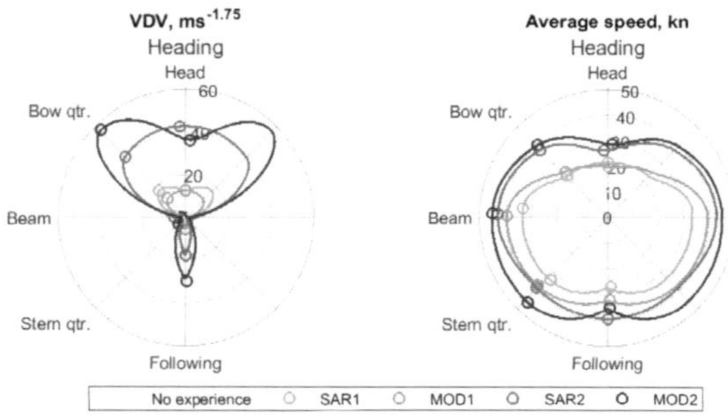

Figure 17: VDV / Transit Speed variation with heading for each coxswain in Sea State 4 (Free/free control)

Figure 15 shows that in SS2, while the transit times are similar for each heading, the overall VDV is dominated by the exposure during either the head or bow quartering seas. This is as expected because these headings involve travelling into and over the waves, and leads to wave impacts when the craft hits a wave or lands in a trough between waves. When travelling at most other headings, the coxswain is able to mitigate large impacts by navigating a path that does not involve such abrupt changes in heave.

Figure 16 shows that in SS3, the highest VDV is generally seen in bow quartering seas rather than head seas. The transit times are also noticeably longer for bow quartering seas, but also for following seas due to the coxswain's trying to avoid the waves overtaking the craft and subsequent yaw/broaching. The overall VDV is dominated by the exposure during either the head or bow quartering seas. The highest overall VDV values are seen where there is also a significant contribution to VDV during following seas.

Figure 17 for SS4 displays similar trends to SS3 except with a much higher VDV and transit time due to the waves being much larger. Again, the overall VDV is dominated by the exposure during either the head or bow quartering seas. The highest VDV values are seen where there is also a significant contribution to VDV during following seas.

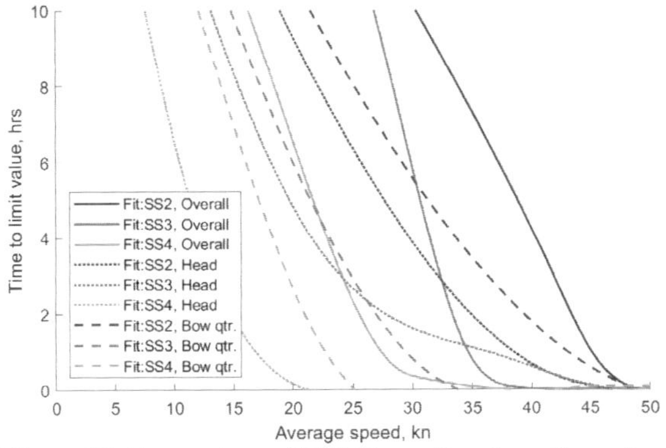

Figure 18: Comparison of TTLV as a function of speed for different headings and the overall TTLV (curve fits only, individual data points not shown)

The trends in TTLV with speed for the two worst headings in terms of VDV (Head / bow quartering seas) are shown in Figure 18. It is apparent that the overall TTLV (solid lines) are not a conservative estimate of the time it takes to reach the limit value if the majority of a mission is conducted in head or bow quartering seas (dash and dash-dot lines). This applies for all the sea states tested. Such performance curves taken in head seas or bow quartering seas could be used to estimate the worst-case TTLV in the event that the heading is not known, or if a mission will be carried out predominately in head or bow quartering seas.

5.3 OVERALL VARIATION WITH CONTROL CONDITION

The results in this Section show how the same key metrics from Section 8.1 compare for each of the control conditions:

A. Free/free – both control types (throttle & steering) are allowed. These are the values already presented in Section 8.1.
B. Throttle only
C. Steering only

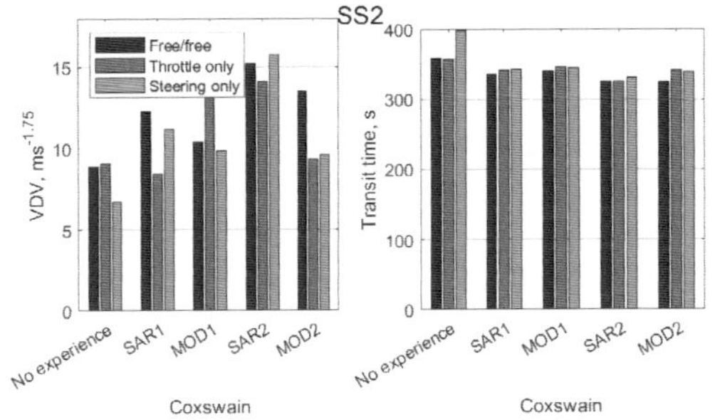

Figure 19: VDV / Transit time variation with control condition for each coxswain in Sea State 2

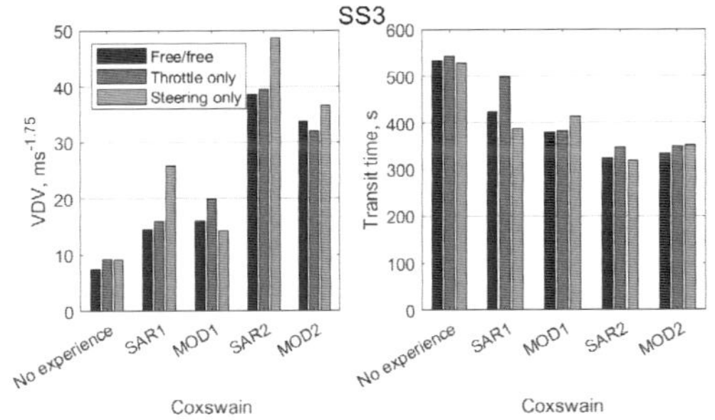

Figure 20: VDV / Transit time variation with control condition for each coxswain in Sea State 3

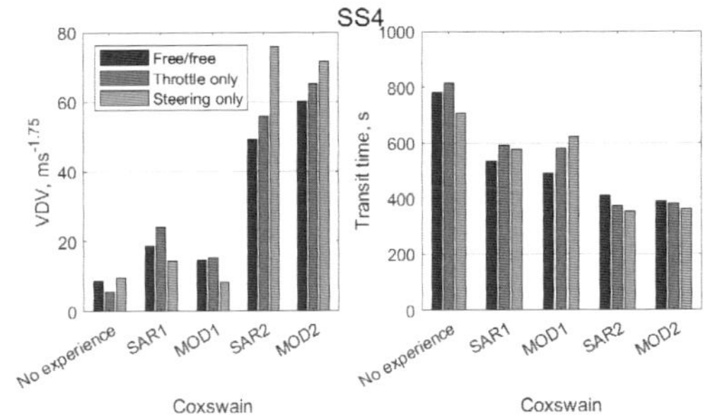

Figure 21: VDV / Transit time variation with control condition for each coxswain in Sea State 4

Looking across all sea states in Figure 19 to 21, there is a general trend of a significant increase in VDV for the steering only case while the transit times remains similar or increase (i.e. transit speed goes down). The trend is clearest in SS3/4 for the SAR2 and MOD2 coxswains who attained the highest speeds. There is also a general increase in VDV for the throttle only case but this appears to be less significant than the increase from having steering only. Figure 22 compares free/free control (solid curve) to throttle/steering only cases for SS3-4. It shows that, for a given transit speed, the lowest WBV is generally obtained under free/free conditions but for a few exceptions at very high or very low transit speed. These results indicate that the throttle is the most important control coxswains use when driving to minimise WBV at higher speeds. This is to be expected as all the experienced coxswains stated in their trial de-briefs that they use the throttle much more than steering to respond to the fast-changing conditions associated with rough sea operation.

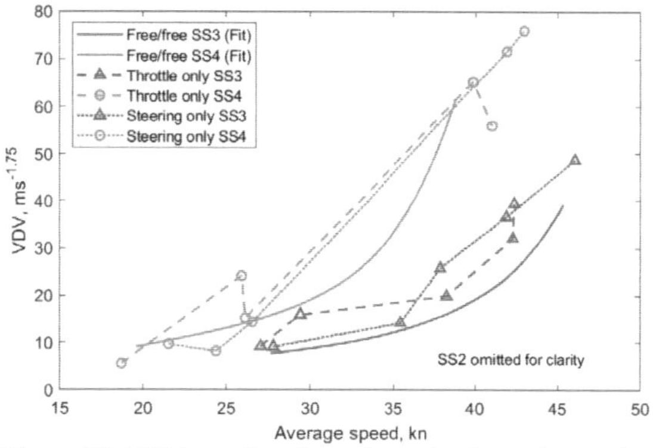

Figure 22: VDV as a function of sustained transit speed across the test pattern for SS3-4. Solid line shows free/free control and dash/dot line shows corresponding throttle or steering only results.

5.5 IMPACT-BY-IMPACT PERFORMANCE CHARACTERISATION

5.5 (a) Impact acceleration correlation parameters with boat speed and wave height or wave slope

For each individual impact, the wave height and wave slope are estimated as discussed in Section 4.11 (d). Nieuwenhuis [1] found that both wave height and slope act as cues for coxswains to moderate craft speed in order to reduce the acceleration peak associated with impacting the sea surface while planing. However, it was to be determined for our approach which of these wave properties correlates most strongly with peak acceleration.

Figure 23 shows peak accelerations recorded in head seas for all coxswains plotted against craft speed multiplied by (a) wave height or (b) wave slope. The inclusion of craft speed in the correlation is important as the peak acceleration clearly depends on how fast the boat is travelling as well. This figure shows that for a broad range of boat speeds (0 – 45 knts) and wave slopes (0 – 12 deg), peak acceleration correlates most strongly with wave slope (R=0.81) compared to wave height (R=0.67). Intuitively this make sense that it is the steepest waves, not just those with the greatest height, that lead to the hardest impacts and therefore highest acceleration peak.

It is apparent from Figure 23 that while there is a strong general trend (see dashed line) in impact acceleration, there is also significant scatter which increases in range with increasing boat speed/wave slope. This range is approximately +/- 2g either side of the general trend. In the remaining Sections, wave slope is used to characterise the severity of the encountered waves and the coxswain's ability to minimise impacts.

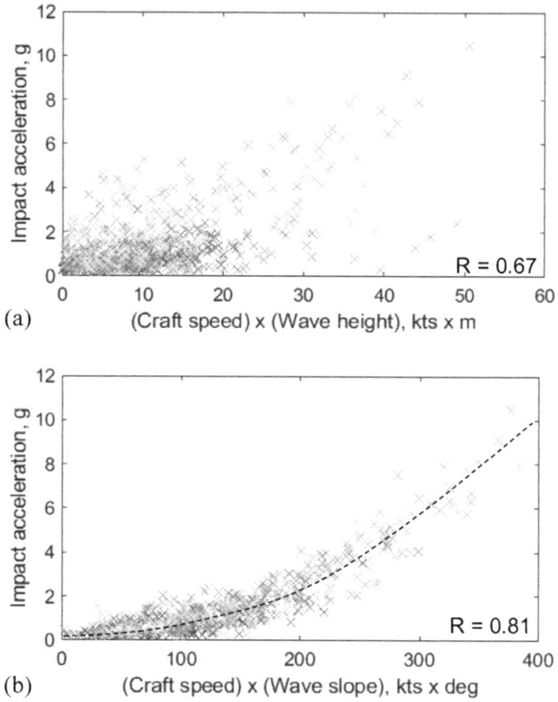

Figure 23: Correlations between impact peak acceleration and (a) Craft Speed x Wave height, compared to (b) Craft Speed x Wave slope. Different shades represent different runs.

5.5 (b) WBV operational maps

Using the same data presented in Figure 23, a performance map for each coxswain can be produced. This shows performance across all impacts in sea states 2-4 as a function of craft speed and wave slope. For the same craft speed and wave slope, there are data points with a range of peak accelerations. The performance map represents the mean peak acceleration across that range.

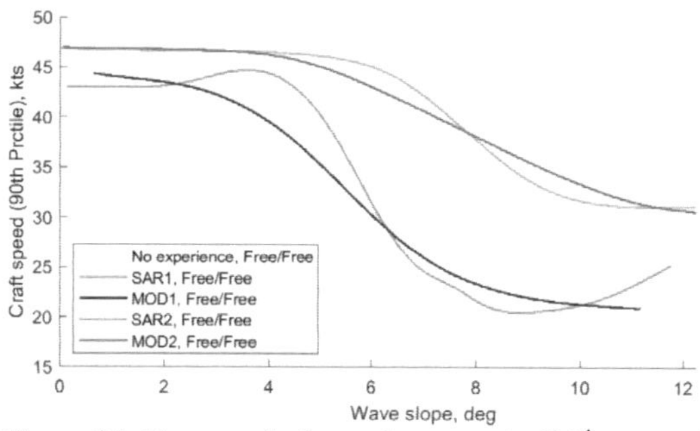

Figure 24: Top speeds for each coxswain (90th percentile) under all free/free control in head seas and sea states 2-4

Typical top speeds for each coxswain are shown in Figure 24. As further illustrated in Figure 25, two of the experienced coxswains (SAR2/MOD2) chose to operate at much higher speeds in steeper waves than the other two (SAR1/MOD1). Operating in top right of the operational map results in peak accelerations in the range 4-8g or more. The other two coxswains avoid operating in this range at all and hence their overall WBV levels are much lower. Additionally, for SAR2/MOD2 at around 35 knts, 8 deg, MOD2 is able to reduce peak accelerations to 4 – 6 g compared with 6 – 8 g or more for SAR2. In other words, for nominally the same incoming conditions, one coxswain reduces peak acceleration by ~10-20% relative to the other through effective use of throttle/steering. A similar advantage can be seen between SAR1/MOD1 at around 25 kts, 9 deg.

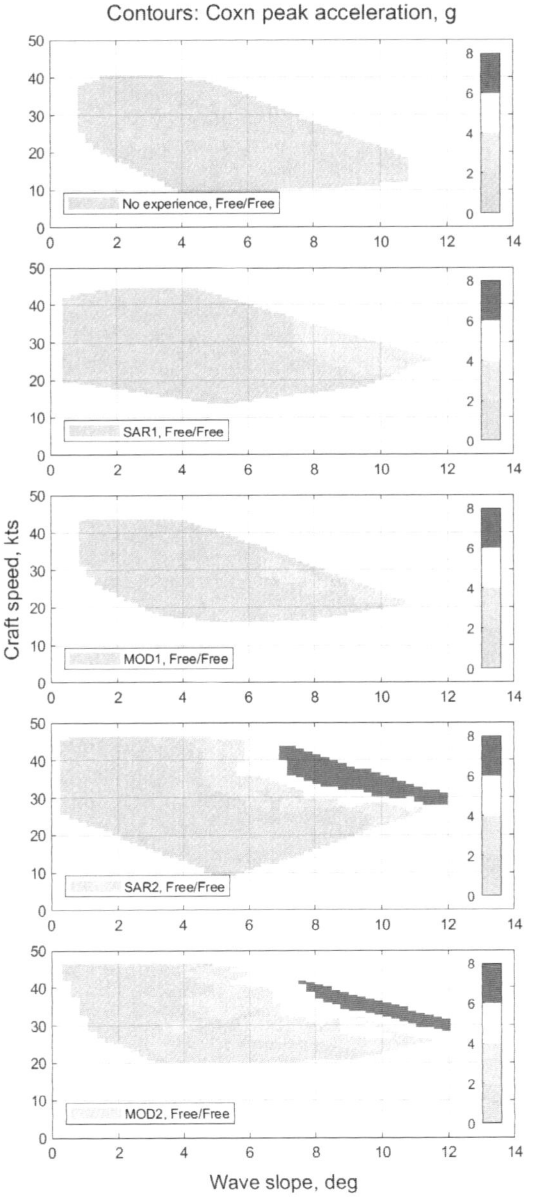

Figure 25: Operational map for each coxswain under free/free control in head seas and sea states 2-4

6. CONCLUSIONS

With reference to the two primary study objectives, the following conclusions are presented.

Objective 1: develop and demonstrate a systematic coxswain behaviour tracking methodology

This objective has been achieved by devising an innovative approach in which a high-fidelity fast craft simulator (developed by MARIN for the Dutch navy) is re-purposed to assess WBV. To the author's knowledge, a simulator approach has not been used before to study WBV and this has been enabled by new simulator technology. The demonstrated advantages of this approach, over the sea trial alternative, have been that it is:

- More repeatable: sea state can be easily fixed such that it presents the same conditions for each coxswain, eliminating the large variations in conditions which are always present in seas trials;
- More controllable: sea state conditions can be varied to give specific conditions within the range of interest.
- Accurate: the simulator been developed by a world-leading maritime research organisation for a similar purpose to our use case and as such is a highly accurate simulation. The trends recorded in WBV exposure compare favourable both in terms of magnitude and range with existing real-world data sources, such as [5].
- More accessible: it does not suffer from the many of the difficulties associated with getting good measurements of quantities such as boat accelerations, wave heights, boat control signals, etc. for multiple participants in a short timeframe.

The potential disadvantages of this approach for tracking coxswain behaviour are that:

- Coxswains may behave differently in studies to how they behave on a real boat as the perception of risk by coxswains will be different to in reality. There will always be coxswains who drive too aggressively because it is considered more of a game. This issue can be somewhat mitigated as long as there are multiple participants; it can actually make it easier to draw out trends in the data with a range of behaviours to compare. In the case of this study, confidence in these trends can be increased through further sea trial work, informed by the simulator trial data.
- Impacts 'hurt less' on a simulator. The peak accelerations are limited by the motion platform and are lower than output by the fast craft model. While this was not a significant issue for this study, if the simulator was used to train less experienced coxswains, it may provide an unrepresentative example to trainees of what happens if they drive too fast. Simulator motion platforms with higher peak accelerations are being developed.
- Some experienced coxswains moderate their speed based on the impacts/pain they feel and are more likely to 'drive too fast' without simulator study controls in place.

As a methodology for tracking coxswain behaviour, on balance, the approach is well suited to WBV studies of this type. Other applications of this approach could be to benchmark coxswains under controlled conditions (*viz* a driving test), to provide systematic feedback on training needs and identify coxswains best suited to execute a certain mission. It can also be used for other human factors research.

Objective 2: quantify the impact of sustained transit speed on shock

This objective has been achieved in a simulated setting and the following key conclusions drawn:

• The experienced coxswains demonstrate are range of behaviour and this is distinct from that of an inexperienced coxswain. Overall there is no single optimum transit speed to reduce WBV but several straightforward trends have been identified that could aid mission planning for WBV reduction.
• A general trend has been established around the benefit of reducing speed on WBV, independent of the sea state (SS2 – 4). On the assumption that VDV is a good indicator of risk associated with WBV, this trend indicates that mission planners could significantly reduce WBV risk to crew with modest (i.e. 5%) reduction in sustained transit speed (equivalent to increasing transit time). This general trend requires further validation using sea trial data. Certain coxswains out-performed or under-performed relative to the general trend by 5-10% which would be a significant reduction in VDV if achieved through increased training and experience alone.
• By breaking-down the contribution to VDV by heading (head seas, bow quartering seas, etc.), the importance of taking into account the prevailing sea conditions into mission planning on WBV has been demonstrated. Overall TTLV curves such as those in Figure 14 or the DIN [1] may be non-conservative. A more accurate approach would be to include curves as a function of the prevailing swell direction.
• By quantifying the relative performance of coxswains when they are not able to control the boat using either the throttle or the steering; effective throttle control has been shown to be much more important than steering to reduce WBV. While this is to be expected based on feedback, it highlights the importance of understanding what constitutes effective throttle control (i.e. what is one coxswain doing that another is not) to reduce WBV.

Further work is required before the findings of this study can be fully exploited. Later phases aim to provide validation and further evidence to support (or potentially contradict) the conclusions outlined above. This work should include a sea trial where coxswain behaviour is tracked in a similar way but under real-world conditions. This verified dataset can also be used to train or test autonomous feedback/control algorithms which aim to minimise potentially damaging impacts.

7. ACKNOWLEDGEMENTS

We would particularly like to thank the professional coxswains who volunteered to participate in the study from UK MOD Marines user groups, the Royal National Lifeboat Institute (RNLI), and the Royal Netherlands Sea Rescue Institution (KNRM).

8. REFERENCES

1. M. Nieuwenhuis, 'Modelling Operator Behaviour - An experimental approach to model operator behaviour on board fast RIBs', *MARTECH / MMT,* D 5050 / 463, July 2005.
2. H. Godwin et al. 'The influence of expertise on maritime driving behaviour', *Applied cognitive psychology*, 27.4 (2013): 483-492.
3. N. Bovens, 'Fast Intercept Simulator', *Maritime Research Institute Netherlands (MARIN)*, presentation, October 2019.
4. P. de Jong, 'FSSS - Hydromechanics model validation - Final Report', *Maritime Research Institute Netherlands (MARIN)*, Report No. 26179-1-SHIP, Ver 1.2, January 2018
5. M. Raeburn, 'Risk Reduction Measures to Counter the Effect from Maritime Whole Body Vibration', *Royal Navy 2016DIN06-016*, July 2016.

9. AUTHORS BIOGRAPHY

Dr Tim Newman holds the current position of Maritime WBV Analyst at the MOD Naval Design Partnering team, a consortium UK Ministry of Defence and the defence industry based in Bristol, UK. He is responsible for Whole-Body Vibration support to the High Speed Craft team and provides expertise for First of Class trials and conducts technical studies with key capabilities around Test & Evaluation (T&E), data analysis and modelling. His previous experience is as a chartered engineer and project manager for a variety of multi-disciplinary, customer focused projects in the defence and wider industry, predominantly on maritime platforms and involving noise & vibration issues and/or fluid dynamics.

LAB TESTING TO DETERMINE SEATS' CAPACITY TO PROTECT USERS FROM IMPACT-INDUCED INJURIES. REBUTTAL OF KEY CLAIMS IN "LAB TESTING OF SUSPENSION SEATS FOR HIGH-SPEED BOATS: LAB TESTING CANNOT DETERMINE SEATS' CAPACITY TO PROTECT USERS FROM IMPACT-INDUCED INJURIES"

PR Moxley, **R Lawrence**, Naval Design Partnering Team, UK

TE Coe, Naval Architect and WBV Analyst, UK

SUMMARY

In recent years, there has been a concerted effort to produce a laboratory standard to test and quantify the performance of shock mitigating seats in high-speed boats. Despite significant progress towards international agreement on the terms of an ISO standard, it was eventually abandoned in favour of varying national-level standards. Some prior publications may lead to misinterpretation, so an overview of the standard is presented to aid the reader in understanding the tenets of the standard, including its objectives and purpose. The remainder of the paper is concerned with rebutting broad categories of objections. Definition of the test waveform, vulnerability of human occupants, effects of filtering on signals, types and suitability of drop test rigs, and the rejection of "brace" type seats. In each instance, examples are provided of where concerns with the test method are unfounded or based on a misinterpretation of the test method.

NOMENCLATURE

A(8)	Average vibration exposure based on example acceleration normalised to an 8-hour work day
BSI	British Standards Institute
Butterworth	A type of signal processing filter
CD	Committee Draft
DEFSTAN	Defence Standard
EuroNCAP	European New Car Assessment Protocol
g	Standard unit for acceleration, equivalent to 9.81 ms^{-2}
Hz	Hertz, unit of frequency
ISO	International Standard
LDTM	Laboratory Drop Test Method
Low-pass	Filter which allows through all energy below a cut-off frequency
MR	Mitigation Ratio
$ms^{-1.75}$	Standard unit for vibration dose value
ms-2	Standard unit for acceleration
NGO	Non-Government Organisation
NDP	Naval Design Partnering Team
RMS	root-mean-square
SAE	Society of Automotive Engineers
SC	Steering Committee
SEAT	Seat Effective Amplitude Transmissibility
Sed8	Static Equivalent Dose (8-hour)
SR	Shock Reduction
SRS	Shock Response Spectrum
TC	Technical Committee
Transmissibility	The ratio of signal output magnitude to signal input magnitude

1. INTRODUCTION

In July 2020, Doctors Ullman, Kasin, and Reynolds, alongside Messrs Eriksson and Borgen presented the findings of an investigation into the now-defunct ISO TC108 SC4 WG 18, tasked with developing an ISO standard for assessing the performance of marine shock mitigating seats. The document produced was circulated as a committee draft, ISO CD 19470 [1]. This was presented as a presentation at the RINA High-Speed Craft Conference, as well as in accompanying article format. The quoted aims of the study were:

- *"...validate the test method, proposed to become an ISO standard, its relevance for tested seats protective capacity, the validity and repeatability of test results and the scientific validity of the claims and assumptions on which it was based.";*
- "*...identify what actual purpose the proposed test standard was intended for and how it was supposed to be used*", and;
- "*Identify an alternative purpose that the proposed test standard could serve*"

Whilst we, as contributing authors to the standard welcome the attempt to provide valid scientific criticism of the proposed test methodology, the resultant analysis misrepresents several key factors about how the test method was to be carried out, the timeline of the project, and the purpose and objective of the method. There are also several glaring processing and mathematical errors contained within the paper.

To correct the record on these points, we aim to present the following:

- A description of the test method as represented in the committee draft produced by WG18 for voting as well as the compromises that were proposed to reach a consensus;

- Corrections of where the science presented by [1] is incorrect or presented without relevant context.

Whilst the original paper makes claims about the motivations of certain individuals and organisations regarding the purpose and motivation of the standard, this document is not concerned with rebutting these. These claims are presented without evidence and are solely the interpretation of the article's original authors. In our opinion, all the contributors to the original CD of 19470 did so in good faith, collaborating openly and working to reach a consensus between NGOs, Government Agencies, and Industry.

To understand the LDTM and the context of both [1] and the current paper, we recommend that readers consider **Section 2, first**.

For those who are either versed in the Laboratory Drop Test Method (LDTM) or are solely interested in the technical rebuttal, we advise starting at **Section 3**.

2. DESCRIPTION OF THE DROP TEST METHODOLOGY

The drop test methodology being discussed here has existed in a variety of forms in the seven years since its inception ([2], [3], [4], [5], [6]), and is anecdotally in use in other countries under various guises. These different versions may be grouped more generally as the Laboratory Drop Test Method (LDTM), and to be fair to the authors of the original article, we will describe the LDTM as outlined in the final ISO Working Group Committee Draft [4].

The full title of [4] is "Laboratory evaluation of marine seat shock isolation", and its final revision prior to the dissolution of the working group was published in Q3 2017. It outlines the requirements for a laboratory or testing house to perform repeatable, vertical-axis drop tests, including; target drop magnitudes, payload masses, the tolerance envelope for accepting an input impulse, and the preferred post-processing methodology. Its purpose is to allow a laboratory or testing house to perform a repeatable test for how much acceleration or energy is removed by the seat relative to the point of input. This value is known as either the Seat Effective Amplitude Transmissibility (SEAT) value, or the Mitigation Ratio (MR).

The first three sections of the document are concerned with a general introduction to the standard. Section 4, "Scope", outlines the formal definition of what the standard is to be used for, making the following statements, which we feel are important to highlight for context:

1. **"This standard applies to seats installed in high-speed marine craft, typically less than 30 m in length"**

2. **"The test method is not intended to assess the seat response to the full range of conditions that may be encountered in practice"**

3. **"The effect of active bracing of the legs or arms is not considered by this laboratory test. [...] fast boat operators have observed less bracing effort occurring as the seat occupants become tired"**

3. **"All other factors, other than vertical shock, that may be relevant to the selection of seats for use in marine craft, such as ergonomics, robustness, materials, restraint systems, performance at different temperatures, and response to non-vertical motions, are beyond the scope of this standard"**

We state these points here, as they demonstrate that the standard is intended for a very specific purpose and that demanding it perform assessments outside of its scope is to fundamentally misunderstand that purpose..

Sections 5 and 6 detail the normative references (i.e. those references of which it is useful to have a working knowledge to properly carry out the tests), and the terms and definitions respectively. The terms and definitions defined in Section 6 are the standard orientation of axes (i.e. the Z- or vertical-axis being parallel to the normal orientation of gravity), and standard abbreviations for vibration terms.

Section 7 is where the majority of the LDTM is properly defined, and we present a summary of it here:

Section 7.1 outlines the definition of the test seat, and what modifications are permissible and necessary to properly conduct the test. It mandates that the seat be a "standard" production model, in no way edited to provide an advantage during testing. Any modifications which are required to rigidly mount the seat to the platform (such as drilling holes into a baseplate) should be carried out in a way which does not alter the operation of the seat relative to how it would be installed on a boat. Adjustments to springs or dampers which are intended to adapt to different occupant mass must be set to the appropriate drop test mass settings, and guidance is given on where to set variable seat and backrest angle settings (in the case of seat height, it is intended that it's height is set to the lowest setting to demonstrate a "worst-case" or highest risk of the seat bottoming out under impact"). Components should be allowed to cool to the ambient temperature of the test environment, and not artificially cooled or heated to alter the performance of said component.

Section 7.2 outlines the payload requirements for the seat. Three payload masses are defined here and are given below.

Light Occupant (M1)	*5th percentile female operator*
Medium Occupant (M2)	*50th percentile male operator*
Heavy Occupant (M3)	*90th percentile male operator*

These masses are taken from anthropometric data from DEFSTAN 00-250 [7] (now superseded by DEFSTAN 00-251) and may be adjusted to consider just the seat-supported mass of the occupant, as well as to accommodate typical kit loads.

Section 7.3 outlines the basic requirements for drop testing without going into fine details, namely that the seat must be dropped under the action of gravity, decelerated to achieve the required impact acceleration, and must be stationary before and after the test.

Section 7.4 outlines the requirements for instrumentation, namely accelerometer performance: ±20 g measurement range (±196.2 ms^{-2}), 0.01 g minimum resolution (0.0981 ms^{-2}), *minimum* bandwidth of 80 Hz (a much higher bandwidth is recommended in the text), triaxial measurement. The placement of the accelerometers is also stated here: one on the base platform to measure the "incoming" vibration, and one on the upper surface of the seat to measure the "out coming" vibration or the vibration passed through the seat to the occupant. The input vibration accelerometer is to be rigidly mounted to the supporting platform, whilst the output accelerometer needs to be an SAE type measurement pad, as defined in ISO 10326-1.

Section 8 is a reference to a normative entry in the annexe describing: some idealised craft classes to be used to determine appropriate magnitudes of shock for a given platform; what test magnitudes are likely to be required for a seat, and; how many impacts are required under the test procedure.

The most nuanced part of the LDTM is Section 9, which details the tolerance envelope of the drop tests. Two acceptance envelopes are defined, one for the vertical axis, and one for the horizontal axis. The vertical axis tolerance envelope is the more exacting, whilst the horizontal envelope is rather broad and intended to identify very unusual off-axis effects. An example of the tolerance envelope is given in Figure 1.

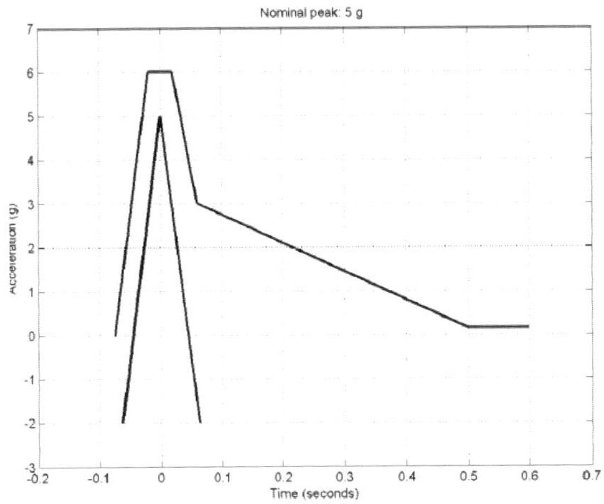

Figure 1 - Example vertical-axis tolerance envelope for a target 5 g acceleration impulse.

The dimensions of the tolerance envelope are given by a set of time and magnitude dependent coordinates and applied to the 20 Hz low-pass-filtered data of the input signal. This is the only pass/fail criteria outlined in the standard, and is intended to allow for repeatable stimuli to be presented both within tests on the same seat, and across different seats in an entire test battery.

Section 10 outlines the requirements for repeated tests on the same seat, requiring that no adjustments be made to the damper, spring, or installation mode between test conditions and that the seat must, as a minimum, be able to complete the test without suffering damage sufficient to prevent its normal operation.

Section 11 defines the test matrix for an individual seat. The requirement is that a total of three valid drops (i.e. drops for which neither vertical nor horizontal acceptance envelopes are violated) be completed at each magnitude x payload combination, to get an averaged output result for each condition. The target magnitudes are defined in an Appendix, but the full gamut is 3 g, 4 g, 5 g, 6 g, 8 g, and 10 g. NOTE: These are the magnitudes of the 20 Hz low-pass filtered acceleration at the point of input to the system. The unfiltered accelerations will be significantly higher than this. A full test matrix then requires 6 impact magnitudes, 3 payload masses, and 3 repetitions, resulting in a minimum of 54 repetitions to produce a full assessment of the seat's performance.

Section 12 outlines the metrics that can be used to determine the performance of a seat. At the time of publication of [4], the preferred metric was the MR of the seat, calculated by the ratio of the Shock Response Spectrum (SRS) at the seat surface and the point of input to the body. The SRS is essentially a mathematical model of the idealised response of the spine to an input signal, based on a mass-spring-damper system with a resonant frequency of 8.42 Hz and damping of 22%. After the publication of [4], the UK and others advocated for the inclusion of other metrics: The UK position that vibration dose value (VDV) was the most appropriate metric to use, as it aligned well with existing legislation, and the consensus was being reached that provision should be made for reporting in multiple metrics, to satisfy all member states. At the time the NDP produced work which showed a good correlation between each of the eventual metrics which was being discussed [8].

The final relevant section is 13, which details what should be reported as part of the output of the test. For clarity, this text is reproduced verbatim below:

"The report shall include, but not be limited to the following:

a. Seat manufacturer, model, description, manual adjustment settings (as tested if applicable)

b. Test date, test laboratory

c. Payload weights, description, method of securing it to the seat and corresponding total occupant masses

d. Description of instrumentation system and sensors including calibration information

e. Description and photograph of seat/payload assembly on test fixture

f. Description of test run-in, execution sequence, and data processing

g. The ambient temperature range during the tests

h. Tabulated SR test results using Table 4 above

i. Example graphs showing seat and deck accelerations as described in Section 12 above

j. Test Class completed and SR_{TOTAL} percentage achieved using the phrasing below

k. Certification statement using the phrasing below.

The following phrasing shall be used in the report for reporting the SR_{TOTAL} percentage for each Test Class completed:

"The overall Shock Reduction percentage (SRTOTAL) obtained for Test Class was %

The best-case Shock Reduction percentage (SR) of %, occurred for mass kg, and at target peak acceleration g.

The worst-case Shock Reduction percentage (SR) of %, occurred for mass kg, and at target peak acceleration g."

The remaining sections of [4] are informative annexes, which attempt to provide broad categorisations of classes of vessel (Annex A), examples of drop test apparatus based on known laboratory resources (Annexes B and C), examples of the nominal 3, 4, 5, 6, 8, and 10 g tolerance envelopes (Appendix D). Example Matlab routine for comparing drop tests against tolerance envelopes is also included (Appendix E).

2.1 PURPOSE OF THE LDTM

This section is intended to provide context for the objectives and use of the LDTM outlined in [4] and [5]. The purpose of the standard is given below:

- It is intended to provide a technical user, such as a researcher or testing house, with a pattern or set of instructions to perform a specific laboratory test;
- It is intended to test the performance, in the vertical axis only, of shock mitigating or shock isolating seats in response to a single, simple impact;
- It is intended to provide a single-figure value for ability to mitigate shock which can be used to quickly compare two or more seats;

The standard is NOT intended to;

- Provide a final filter to determine whether a seat is safe to take to sea without further testing;
- Provide a numerical assessment of a seats' capacity to reduce injury;
- Remove any legal burden from fleet operators in their duty to protect workers from shock and vibration;
- Test the at-sea performance of a shock-mitigating seat.

3. REBUTTAL OF TECHNICAL CLAIMS

Now that we have provided the reader with the basic premise of the drop test methodology as outlined in [4], we will address the technical concerns raised in [1]. To present these rebuttals, we have grouped these claims as necessary to avoid repetition and presented the claims from [1] for context.

3.1 DEFINITION OF IMPACT WAVEFORMS

3.1.1 "A 100ms half sinus [sic] *wave is not representative of and does not simulate a slamming induced impact as measured at the hull of a high-speed boat. These do not even remotely resemble a half sine wave." (Section 3.3, p.21)*

The authors of [1] claim that the incoming waveform is defined by the standard as a "half sinus" [sic] waveform. The phrase used in [4] is "nominal half-sine impulse". This refers specifically to how a drop test can be identified and standardised when viewed as a 20 Hz low-pass filtered signal. It does not mean, as the authors of [1] intimate, that the drop test method is intended to produce a pure half-sine waveform. For further clarification, an example waveform is presented alongside the tolerance envelope in Figure 2 (note that, in line with [4], this is 20 Hz filtered data)

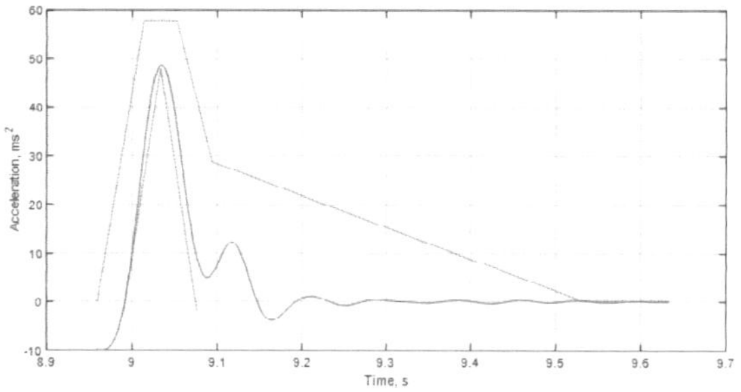

Figure 2 - Example drop test waveform compared with vertical-axis tolerance envelope. Waveform filtered at 20 Hz

The task of deriving the initial drop-test tolerance envelope from real-world data was given to the Human Factors Research Unit at the Institute of Sound Human Factor Research Unit, University of Southampton. This initial work was based on a single acceleration time history recorded on a high-speed craft in rough weather and developed an envelope based on one shock magnitude, 7g. In order to generalise this approach further studies were undertaken to validate the envelope across a range of shock magnitudes by defining typical wave impact features.

As part of the work done to define a "typical" impact it was necessary to define what the duration of such an event was - the difficulty with this is that an impact recorded at sea is part of a broader set of acceleration events at varying frequencies. As such, picking out impacts from the associated background noise is difficult. To better define these events, half-sine curves were fitted to the identified impacts that allowed their duration to be assessed based on the zero-crossing period of this modelled sinusoid. The suitability of this modelling approach is further supported by [9], which analyses wave slam events greater than 2.0 g to determine their typical characteristics. This work was independently validated in [10], which outlines requirements for simulating wave impact events based on the work outlined in [9]. To wit; *"Review of acceleration data recorded during full-scale seakeeping trials demonstrates that wave impact durations can vary from about 0.1* (100 ms) *seconds to 0.4 seconds* [400 ms] *... the shorter durations tend to occur for the more severe wave impacts, and longer durations are observed for less severe impacts and heavier craft".*

Having developed these envelopes algorithms were developed to compare them with a large number of impacts recorded at sea and the envelope was found to be a reasonable representation of real-world impacts across several craft.

While durations considerably greater than 100ms can be recorded at sea as shown in [10] and fitted in the envelope these represent a greater challenge to mitigate. The 100ms minimum duration, however, appears to be a clear limit. Due to the nature of a single degree of freedom isolator (such as a marine shock mitigating seat), it is easier to isolate shorter-duration impacts. Therefore, it was felt important to ensure that unrealistically short impacts were not permitted especially if manufacturers were self-certifying seat performance.

3.1.2 "Seats need to be tested only for typical and not for worst-case impacts. These claims are false. Numerous user agencies have documented hull impact exposure levels exceeding 20g. [...] Seats approved for use by such lab testing, but not tested for real-life exposure levels may give a false feeling of security and hence cause increased risks of injury. (Section 3.1, p.20)"

The specific text from the article attributes the following statements or sentiments to an unspecified ISO WG member: *"Typical slamming induced impacts on hulls can go only up to 10 g and impacts above 10 g are 'a one in a million years event'"*, and states that "*Performance only to 10 g is questionable".* This comment has three logical flaws, namely, data bandwidth, actual distribution of peak magnitudes found in sea trials and the approaches of other safety tests. Considering bandwidth in the first instance what the authors of [1] fail to consider is that this value of 10 g refers to the 20 Hz low-pass filtered data, not to the raw, peak acceleration of the impact. Removal of the high frequency vibration components to leave the rigid body motions of the boat is necessary to compare the at sea measurements with the LDTM [11]. The authors of [1] themselves show exactly this process in "Graph 5" of their paper, which is recreated in Figure 3 for clarity, although it should be noted that they appear to have applied the filter incorrectly.

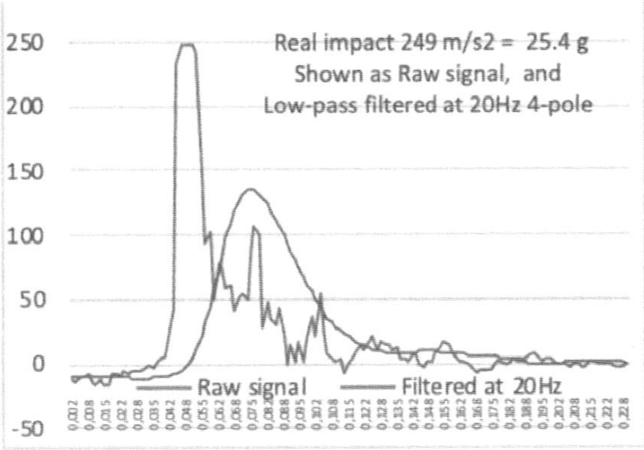

Figure 3 - Figure recreated from [1]

To demonstrate the expected behaviour of a correctly-applied 20 Hz and 10 Hz 4-pole low-pass Butterworth filter, the peak event presented in [1] was recreated and filtered accordingly. The results from this are presented in Figure 3.

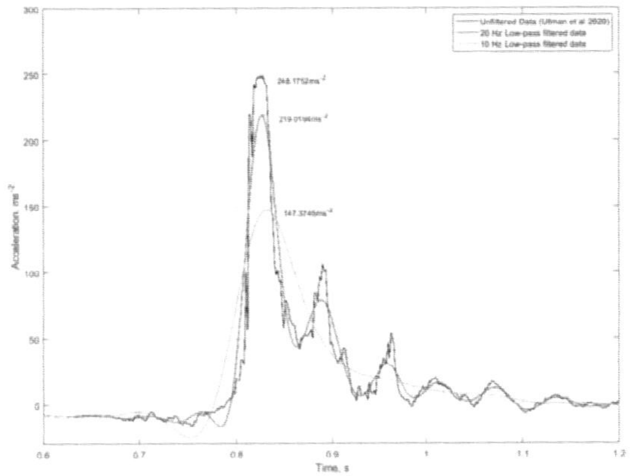

Figure 4 - Example impacted recreated from [1], alongside 20 Hz LP filtered (red) and 10 Hz LP filtered versions (green) of the same impact

It should be noted that a) the data presented by the authors of [1] is clipped at around 25 g and b) the filtering they have used is neither what the standard suggests, nor what the authors claim it to be. We have not been able to recreate the exact way in which this data is processed, but it is not a 20 Hz low-pass 4-pole Butterworth filter. It may instead have a cut-off frequency of less than 10 Hz (which would explain why so many features have been removed from the trace), as well as not being a zero-phase-lag filter implementation, which produces a significantly time-shifted peak compared to the input. This may explain some of the comments regarding both waveform shape and lack of repeatability.

Taking the second point, the authors are correct in their assertion that the LDTM does not test to the highest possible impact magnitudes. There are however several reasons for this approach. Firstly, wave impacts have been found to follow a long tailed distribution [12] where the upper bound is not fixed but where large impacts have a very low probability of occurrence. It is therefore impossible to define a maximum acceleration magnitude. In addition they neglect to mention that recreating larger magnitudes in a safe and repeatable way is incredibly difficult. The drop test rig used by the NDP (located at the RAF Centre for Aviation Medicine) has a maximum drop height of around 2.5m and a total height of around 3.5m and is therefore incapable of producing the exceedingly rare magnitudes seen in the worst instances. It is also broadly representative of the sort of capabilities of other drop test apparatus around the world. For a standard to be of use to anyone, it must be achievable. For a laboratory standard to mandate levels over and above what any laboratory in the world can reproduce would be a pointless exercise. The insistence that higher impact magnitudes be achieved has usually accompanied a lack of understanding of the place of the laboratory test in the overall procurement process.

Finally, the authors of [1] suggest that this approach is at odds with that seen elsewhere in industry however even a cursory search of such test protocols shows that it is entirely normal to test at some practical maximum below the absolute worst case that can be achieved.

To summarize the choice of 10g as a maximum impact magnitude is based on:

1. It is a severe use case. Wave impacts are distributed statistically, following a Pareto distribution [12] and, as such, large impacts are unlikely but not impossible.
2. It is practically achievable with drop test rigs of less than two stories total height while maintaining the minimum impulse duration.
3. It follows the approach used in other testing protocols, for example, EuroNCAP vehicle crash testing [13] where the majority of tests are carried out at 32, 50 and 60 km/h. Rather than the maximum achievable speed of the vehicle.

3.2 VULNERABILITY OF HUMAN OCCUPANTS TO LATERAL IMPACT COMPONENTS VS VERTICAL IMPACT COMPONENTS.

3.2.1 "It is well established that the human spine is more vulnerable to impact forces in the horizontal plane than it is to purely vertical" (Section 3.2, p.21)

The claim here is that it has been '*well established*' that the spine is more vulnerable to impact forces in the horizontal plane than it is to purely vertical, but this claim is presented without any supporting evidence. If the authors themselves have established this, then they have not presented any of their evidence in either the ISO working group or [1]. We have found little to no evidence of lateral shock being the most injurious component of the marine shock environment [14], and to state this to be the case without providing supplementary evidence is not good practice for scientific work. We accept, as "common sense", that lateral shock components are important to human exposure and injury mechanics, but not that they are the most important or the key driver.

Further to this, as a point of clarification, it should be noted that the claim made by the authors of [1] regarding the apparent denouncement of testing for non-vertical impacts components is attributed to comments made by Dr Coe (referenced as "Tom Coe at WG Meeting"). Dr Coe would like it known that he does not recognise this statement as attributable to him, and wishes to issue the following clarification of his position:

"With this standard we are addressing vertical impacts and we completely agree that 'non-vertical impacts', or rather, 'vertical + non-vertical' impacts are potentially a bigger issue. However, this is not within the scope of this standard".

The UK Royal Navy carried out extensive surveys on their fleet of high-speed craft from 2011-2013 measuring triaxial accelerations in a range of conditions. These data showed that in the vast majority of cases (81%) the vertical vibration exposure was higher.

As a final point to this argument, it should be considered that, at the time of writing, there are few commercially available seats containing a mechanism for reducing the impact of lateral shock. and to the best of our ability to determine none that have provided evidence of their capacity to do so. Seats are typically designed to function in the vertical axis only, with very few exceptions, meaning that the specific apparatus that makes them "shock mitigating" is constrained to work in the vertical axis, and typically provides little to no mitigation in any other axis. Typically, where seats can provide mitigation in the lateral axis is through the inclusion of seatbelts, harnesses, lap-belts, and seat bolsters which partially constrain the motion of the upper body. The LDTM is intended to test the vertical performance of the seat only, meaning, specifically, its shock mitigating components. The performance of any other aspect of the seat, whether this is lateral shock performance, aesthetics, cost, durability, etc., are specifically outside the scope of the method and are left to the procuring agents to test using other means. This will be addressed fully later on in this paper, but is stated here for emphasis: The LDTM is not intended as a pass/fail criterion for the purchase of seats. It is clearly stated to be a laboratory standard for the testing of a specific aspect of the performance of a shock mitigating seat to aid in procurement decisions

3.2.2 "It is most probably stemming from observations that cadaver specimens of lumbar spines seem to show an inherent resonance tendency around these frequencies" (Section 3.4, p.22)

In section 3.4 of their paper, the authors of [1] make specific reference to the LDTM being based on an incomplete understanding of the behaviour of a cadaverous spinal lumbar section. This is only partially correct: during the ISO working group, several options were discussed for the reporting metrics from the LTDM, ranging from peak acceleration to vibration dose value (VDV), to shock response spectrum (SRS), and "Static Equivalent Dose, 8 hrs" (SED8). Whilst SRS is a more general vibration term which relates the response of a theoretical, parameter-tuned, single degree of freedom system to an impulse input, it's application to seat testing is specifically tuned to best represent the response of the upper body of the seated human. The particular response for human occupants was formulated in [15], and it is the values presented there that form the basis of the SRS approach to reporting seat performance in [4].

Whilst the final version of [4] included the option for testing houses to use any of these metrics, depending on the requirements of the region, or of the body requesting the testing, early drafts did specify the SRS value. Through our own research as part of efforts to produce a standard that was acceptable to all member nations, we have shown that VDV, SRS, and SED8 all produce results that are well correlated with one another.

3.3 GENERAL CONCERNS AROUND FILTERING AND "TRANSFORMING" DATA

3.3.1 "VDV, Vibration dose value is developed to quantify the mean level of energy transmitted as vibration. This unit does not correlate to the actual forces of discrete impacts nor to any physiologic response Sed8 has the same limitations."

A recurring theme in our interactions with the authors of [1], and repeated in [1] itself, is that by transforming data away from raw acceleration values, the authors of the LDTM are somehow attempting to delude or mislead the non-technical reader. The most frequent objection is that the low-pass filtering of data "removes" key elements of the waveform from the analysis undertaken and that there are still significant contributions to injury mechanics over and above a 20 Hz cut off frequency. There are, to the current author's way of thinking, two misconceptions tied up in this argument, as well as a marked refusal to assess the evidence presented during the ISO working group period.

The authors of [1] are correct that there is 20 Hz low-pass filtering involved in the LDTM, and it occurs at two different stages and depends partly on which measurement metric is required for the testing. The first aspect of low-pass filtering occurs at the "drop-test acceptance" stage of the method - once a seat has been subjected to an impact, the input acceleration, measured on the test platform, and is low-pass filtered at 20 Hz so that it can be compared to the tolerance envelope. Let us refer to this as "tolerance filtering". The second instance of 20 Hz low-pass filtering occurs if the results from the test are being reported using the SRS methodology, let us refer to this as "SRS filtering".

As far as "tolerance filtering" is concerned, the 20 Hz low-pass filtered data is effectively discarded as soon as it has been established that the tolerance envelope has been met, and this is applied only to the input acceleration measurements below the seat. The unfiltered data at both the seat surface and the point of input to the seat are then processed using the preferred metric of the laboratory, testing house, or purchasing body.

"SRS filtering", as stated previously, does involve low-pass filtering of the data at 20 Hz, which, for data sampled at a rate of 512 Hz, excludes energy at frequencies between 20 Hz and 256 Hz. To those not used to working with this type of data, this may seem like a huge exclusion of data, but this ignores both the frequency content of the input signal and the biodynamics of the human body.

In practice there is very little energy contained in boat acceleration time histories in these frequencies [11] so although there is a lot of data at these high frequencies the bulk of it represents low amplitude vibration (such as the flexure of deck panels excited by the engine).

The preponderance of research in the field of human response to shock and vibration pertains to the effect of continuous vibration on the human body, rather than impulsive vibration, but there are still valuable lessons to be learned from this data. The response of the human body to vibration is usually classified by the "apparent mass" or "transmissibility" of the human body. A good overview of the calculation of these properties is given in [6], but their full definition is outside of the scope of this article. In simple terms, the apparent mass is a transfer function derived from the comparison of the input force at frequency F(f) and the output acceleration measured at a given point on the human body at the same frequency a(f). This is often normalised by occupant mass to derive the "normalised apparent mass", which better facilitates comparison between subjects of different masses. "Transmissibility" is defined similarly, but is instead concerned directly with the transfer of acceleration at the point of input to the human body to some other point in the human body (seat-to-head transmissibility, for example, is the transfer function between the seat and the head). It is not in anyone's best interests to present an exhaustive review of the biodynamics of the human body here. Instead, we defer to the work undertaken by [17] that reviewed relevant studies presenting transmissibility data. Their results have been replicated here for visibility in figures 5 - 7. For the non-technical user, the y-axis interpretation should be considered as follows:

- **Transmissibility < 1.0**: acceleration at this frequency is reduced between the point of input and the point of measurement;
- **Transmissibility = 0.0:** acceleration at this frequency is neither amplified nor reduced between the point of input and the point of measurement;
- **Transmissibility > 0.0:** acceleration at this frequency is amplified between the point of input to the body and the point of measurement.

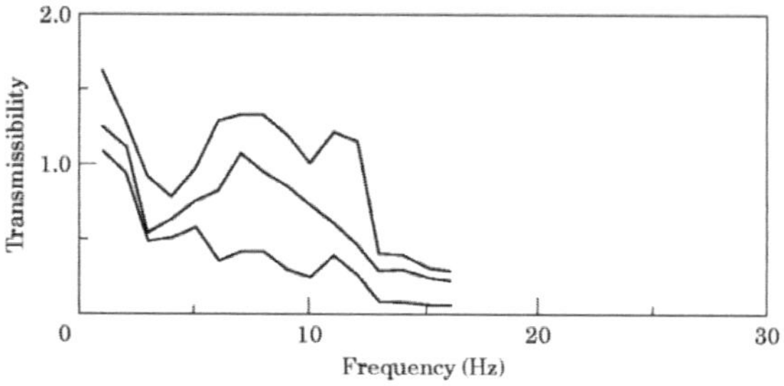

Figure 5 - Median and interquartile values from a review of data from ten studies showing averaged fore-and-aft seat to head transmissibility, reproduced from [17]

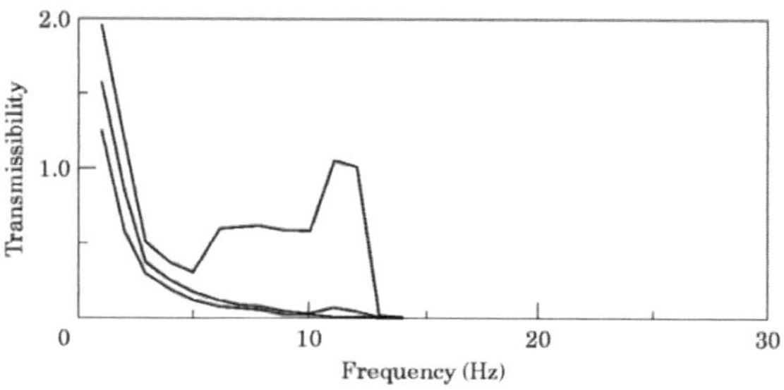

Figure 6 - Median and interquartile values from a review of data from 14 studies showing averaged lateral seat to head transmissibility, reproduced from [17]

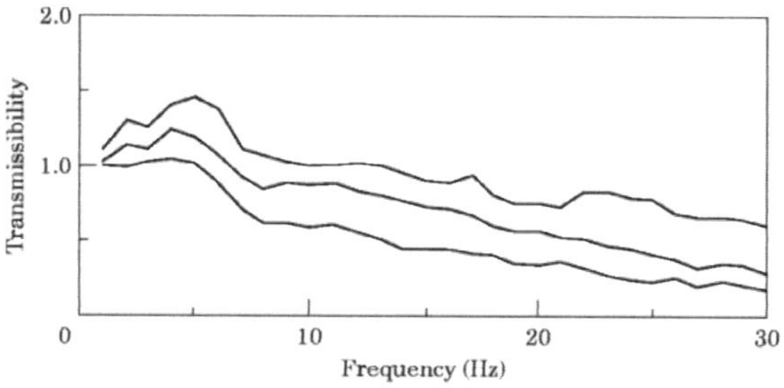

Figure 7 - Median and interquartile values from a review of data from 46 studies showing averaged vertical seat to head transmissibility, reproduced from [17]

For each of the three translational axes of vibration, the review shows that the human body has its peak in transmissibility below 20 Hz. For lateral and fore-and-aft vibration, transmissibility typically falls off sharply between 10 and 15 Hz, particularly when using the upper quartile measurement. The roll off for vertical vibration is significantly slower, but even when using the upper quartile measurement, the transmissibility of the human body from the seat to head is significantly less than one by 20 Hz.

Two corollaries are raised here for transparency: firstly, the human body is a highly non-linear system, and its performance can vary significantly with the magnitude of the stimulus, and secondly; the data from [17] is concerned with continuous sinusoidal vibration, not shock. On this first point, several studies demonstrate how the resonant behaviour of the human body alters with input magnitude. By way of example, [18] shows that, for sinusoidal vibration, increasing the exposure magnitude decreases the resonant frequency of the human body for measurements at varying positions along the back in the vertical axis. To the best of our ability to determine otherwise, there is no evidence that the non-linearity of the human body meaningfully increases the transmissibility of the human body at frequencies above 10 Hz.

Concerning the lack of data for the biodynamic response to shock, or continuous vibration containing shock, we can only present what evidence is readily available to us. The most relevant work undertaken on the subject does not address the biodynamics of the human body, but rather considers the relative discomfort of shocks input at different frequencies [19]. The study shows quite clearly that there is a marked decrease in sensitivity to shocks with increasing frequency (or, perhaps more appropriately for shock, with decreasing pulse duration). We do not claim that vibration sensitivity and risk of injury are exactly correlated, but this is one of the few pieces of robust evidence that exists with which we can start to assess what the likely sensitivity is for human occupants as a function of frequency.

3.3.2 "Most important however, is that there is no reason to believe that singular impacts cause any resonance at all, especially none that would keep resonating after a second, when the next impact could be expected to occur". (Section 3.4, p.22)

This statement is particularly troubling because it demonstrates a lack of understanding of basic biodynamic nomenclature and a basic contradiction in the general thrust of the article. Firstly, the term "resonance" does not simply mean that one expects to strike a structure and have it ring like a bell. Rather, it demonstrates where a system is most pliable, the frequency at which one can expect the largest response of the system. Secondly, the idea that impacting the human body once does not produce any kind of resonance response is easily dismissed as incorrect by considering any amount of published research on the subject.

3.3.3 "VDV, Vibration dose value is developed to quantify the mean level of energy transmitted as vibration. This unit does not correlate to the actual forces of discrete impacts nor any physiologic response Sed8 has the same limitations."

It is not accurate to say that VDV is a measure of the mean or average energy in a signal. VDV is more correctly defined as a measure of the <u>total energy</u> in a signal. The formula for VDV is given below, taken from [20]:

$$VDV = (\int a_w^4 \, dt)^{1/4}$$

Where a_w is the frequency-weighted acceleration time history in ms^{-2}. The presence of the "w" subscript denotes that the acceleration time history should first be weighted by the appropriate ISO standard frequency weighting from [5]. The 4th power term is an artefact of its original intended use - evaluating vibration exposure in environments with a high crest factor (ratio of peak acceleration to RMS acceleration), as it effectively "weights" the impact of peak events so that they are not lost in the noise of other measures such as A(8) or RMS acceleration. As indicated by the integral term, this is functionally distinct from a "mean level of energy" and instead represents the total energy in the system.

The second place where the concept of VDV as it applies to drop testing is discussed in [1] is in "Graph 6", which is recreated in Figure 8 for clarity.

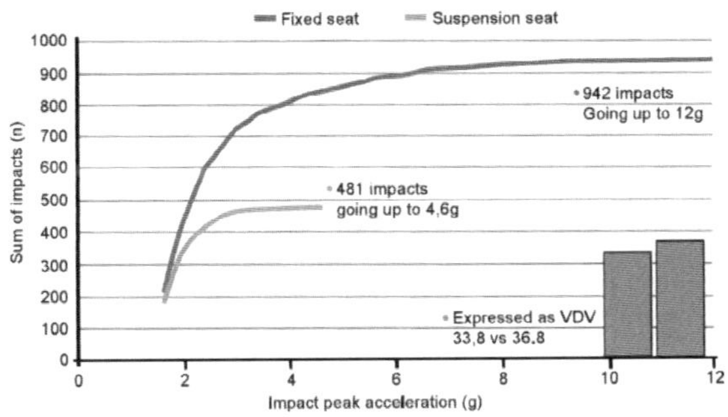

Figure 8 - "Graph 6" from [1], reproduced here for clarity. Cumulative distribution of impacts on a fixed and suspension seat, alongside bar charts for VDV of the same

The data presented above relies on a methodology presented by [21] that effectively counts the number of impacts above a given threshold. We will not go into the merits or demerits of this methodology here other than to say it may be overly simplistic. Rather, we wish to address how the VDV value for the two conditions (fixed seat, suspension seat) has been produced. In the description of the underlying dataset, the authors of [1] state that this is data taken from a three-hour transit on a military patrol boat, and we judge that the VDV data presented is for that entire three-hour transit. The argument being made is that, because VDV does not produce as large a difference in this 'three hour' setting as a simple count of the number and peak acceleration of impacts, it is unsuited for use in a laboratory test. However, this is not an appropriate comparison and constitutes a gross misunderstanding of how measuring the VDV of a single impact differs from measuring the VDV of a three-hour dataset.

In personal correspondence with one of the authors of [21], we were able to determine that the peak-counting method makes the same essential simplification as using the "peak" acceleration to determine seat performance - the peak events contain energy at all frequencies, and subsequently contains a great deal of energy which is not significantly correlated with human body effects. The author's summary of the method is that it was intended to produce a result that maximised inter-seat difference, but is of little to no use in determining the ability of a seat to protect a human occupant.

The purpose of a shock mitigating seat is to reduce the magnitude of shock transmitted from the deck of the boat into the user - barring some effect of the cushion, shock mitigating seats, as a rule, do little to mitigate continuous vibration or shocks below around 1 g. In a three-hour measurement period, the majority of energy into the seat system comes from low-level shock and continuous vibration, not from the kind of large impacts they are designed to protect against. What the authors of [1] fail to present here is that the way VDV, SRS, and SED8 are used in the LDTM is to assess a single large impact, the kind that these seats are designed to protect against. Even in this instance, the reporting specified by the LDTM does not use VDV, SRS, or Sed8 as the output; it uses the SEAT or MR values that being the ratio of these metrics at the point of input to the seat surface.

One of the consistent strains of thought throughout [1] is that its authors purport to be concerned that the metrics used to determine seat performance are too simplistic, or that they exclude too much energy due to filtering etc. However, the two alternative approaches outlined by [1] are, if anything, even more simplistic.: The authors variously advocate for peak acceleration, and impact counting index, these being, respectively: a single figure referring to the maximum acceleration recorded during a slamming event, and; the number of peak events encountered above an arbitrary threshold.

3.4 THE USE OF SAND, LACK OF SPECIFICATION OF THE DROP-TEST RIG, LOW REPEATABILITY.

3.3.1 "The proposed standard does not prescribe a specific method for absorbing the impact of the test platform while impacting." (Section 3.7, p.24);

A fundamental underpinning of the LDTM is that it does not mandate a specific type of drop test rig, recognising that the definition of a bespoke drop test system was beyond the scope of the LDTM. Instead, two informative annexes are provided which show examples from contributing ISO working group members; one using a wedge impacting a crate of dry sand to generate the shock, the other using aluminium honeycomb tubes. As [1]'s authors' principle interaction has been with the UK MoD's NDP, who utilise the sandbox method, this is where the majority of their criticism is levelled.

As there is no mandate in the standard to use a sandbox, we do not feel it necessary to offer a point-by-point rebuttal here. If the authors of [1] do not wish to use a sandbox, they are not obligated to. Instead, the standard provides acceleration tolerance envelopes that any drop test system must reliably be able to recreate. One point that we do wish to address, however, is the claim of a 31% deviation in measured impact. It is difficult to guess exactly what the authors mean by this, as they present no data and no references, and have not produced a meaningful volume of drop test data for us to examine. This level of variation would suggest to us that the drops achieved by them did not meet the criteria of passing the tolerance envelope. We would be happy to assist the authors with the design and interpretation of their drop test data and would welcome greater clarity as to where they derived this error rate. The NDP remains one of the worlds most experienced users of the LDTM and has never experienced error rates like this.

3.3.2 "...Impact testing that uses a perishable material to obtain retardation, is less than ideal, as this entails the need for all test facilities in the world to use the same material, from the one and same manufacturer for all testing (Section 3.8, p.24)

The concerns raised here are based on a misinterpretation of [4], itemised as follows:

1) It asserts that a sandbox is mandated by the standard when it is not;

2) It makes un-evidenced claims about how the varying properties of sand significantly affect the drop test method, without considering that the tolerance envelope (which is mandated by the standard) puts specific repeatability criteria into the method;

3) Suggests that wear and tear on sand particles is a significant issue;

4) That any user of the LDTM would not reset sand after each impact;

5) That tests performed on sand that has not been reset to a level position would not violate the tolerance criteria. We reject all of these assertions out of hand as not being based on a correct interpretation of the LDTM. The use of sand is not mandated by the standard, it is simply presented as an example that has been shown to work well.

3.3.3 "With this method, test results will be significantly influenced by a number of interacting factors such as ..." (Section 3.6, p.24)

In this section, the authors present a list of factors that they believe will significantly affect the outcome of any drop testing. However, the factors they list are either: part of the design of the seat being tested, or; specifically controlled for within the text of [4]. These are categorised below, alongside our categorisation of these factors.

Intrinsic to seat design:

- The pre-tension of the spring element;
- The rebound resistance of the damper;
- The resilience characteristics and dimensions of the seat cushioning;
- The stroke of the seat suspensions.

Explicitly controlled by the LDTM:

- The elastic properties of the dead weight representing the human occupant;
- How the sensor is attached to the top of the padding;
- The duration of the platform deceleration.

Implied control via experienced professionals:

- The slack in fixation and actual relative movement of the dead weight
- How the sensor is attached to the top of the padding
- The elastic properties of the dead weight representing the human occupant - hard or soft mass
- Weight of the suspended part of the seat relative to the weight of the test platform including the unsuspended part of the seat

Where something is listed as "intrinsic to seat design", the authors of [1] are correct to say that this can cause differences in measurement between seats, which is to be expected. If you take a seat with a 10 kg suspended payload mass on a very soft spring, this will have a very different response to a seat with a 100 kg mass on a very stiff spring. We leave it to the reader to determine if the fact that the LDTM will give a different response for seats with a different design is a shortcoming.

Where something is listed as "Explicitly controlled by the LDTM specification", they are just that – not "uncontrolled" variables, but aspects of the test that are acknowledged as having the potential for variability. Consequently, they are controlled to the best of the standard's ability to do so. If it is the view of the authors of [1] that they were insufficiently controlled, then the ISO working group would have been the forum to discuss this and make suggestions for how guidelines may have been improved.

Where something is listed as "Implied control via experienced professionals", it is our view that the point raised can be dismissed by ensuring that the LDTM is carried out by a competent and experienced test and instrumentation organisation, who are aware of the intricacies and difficulties associated with shock and vibration measurement. Given that the authors of [1] have professed to believe that the LDTM mandates that seat manufacturers carry out the testing themselves, we understand why they may be concerned, but as outlined earlier in the report, this is not the case. Any instrumentation and testing company/facility should be selected for this task on the back of the due diligence of the procuring body - a laboratory test standard cannot mandate this kind of due diligence.

We do acknowledge the comments of the authors of [1] regarding the mass with which the seat is loaded. It is necessary for the seat to perform when loaded in the same manner as when occupied by a person. As discussed above a human occupant cannot be considered a rigid mass however studies carried out by the Institute of Sound and Vibration Research at Southampton University have shown that under boat like motions a rigid mass produces very similar seat motions to a human occupant. It would be an improvement to the process to use an object with similar dynamic responses to a human, that however limits the experimenter to a choice of 2 options, the first being live humans with the ethical considerations that entails or a dummy with similar properties to a human. Dynamic dummies do exist such as the ISVR HVLabs dummy however they are not suitable for acceleration levels above about 9.81 ms^{-2} . Attempts to create an agreed upon standard for such dummies continue.

NOTE - The authors have not come across any research where human subjects have been subjected to repeated shocks of the magnitudes seen at sea under laboratory conditions. It is not considered a normal ethical procedure.

3.5 REJECTION OF "BRACE TYPE" SEATING

3.5.1 "...solid scientific evidence indicates the opposite. Scientific evidence exists showing that some such seats, optimised for synergising with muscular reflex response, without the use of any bracing, can reduce physical fatigue by 31% and impact exposure by 50% in numbers and 63% in peak levels."

The method detailed in [4] does not exclude the testing of this type of seat - it excludes the testing of this kind of seat with an active human occupant who can use the seat as intended. A seat that allows the user to brace is not a unique design, and is the default for any jockey style seat on which the user sits with their feet on the deck of the boat. All LDTM seats are tested without the human action, which puts the seats on a level playing field, and removes a huge variability in occupant mass and applied force. It is recognised that the legs can take an active role in shock mitigation, however, the LTDM take the view that ignoring this is a worst-case/conservative scenario for when seat occupants are not using their legs actively. Effectively, the issue of using systems outside of the design of the seat - i.e. using the legs - is not within the scope of the LDTM.

We are aware, broadly, that [1]'s authors are concerned that their seats in particular are unfairly represented in drop testing as the human aspect of the system is excluded, and as such their actual mitigation is higher in real-world settings than in drop testing. The seats produced by the authors of [1] do not comprise a unique design centred on *only* working when human muscle activity is included - the inclusion of human action is so common to shock mitigating seats for use in high-speed craft as to be almost considered the default design. Ergo, when these seats are evaluated for their ability to mitigate shock using the LDTM, they are tested on a level playing field with all other seats of the same design. If the seat was to exhibit better performance with a human occupant, then so would every other seat designed the same way. Conducting the test like this does not place any specific manufacturer at a competitive disadvantage.

4. SUMMARY

Excluding criticisms about the organisation of the ISO working group, misrepresentations of member motives, and the overall timeline of the standardisation process, the authors of [1] have technical complaints centred on four main themes:

- The definition of the impact waveform;
- An appropriate representation of the vulnerability of the human body;
- Filtering and appropriate metrics for reporting shock;
- The use of specific drop test methods, and concerns around the variability of the drop test method.

Through the course of this document, we have highlighted where these claims are either; not supported by the available evidence; based on a misinterpretation of the text of [4], or are factually incorrect. It is our view that the paper, as published, lacked scientific rigour in that it makes assertions without providing robust evidence in the form of example data or appropriate references. Further to this, through communications with Dr Reynolds, we have established that at least one of the authors listed as being involved in its writing was unaware of its existence or publication.

The LDTM is a simplification of real-world events and could doubtless be improved upon with further effort. However by assembling a robust, controllable and repeatable method backed up by published data, it is an effective tool for measuring vertical shock mitigation performance as part of determining a suspension seat's fitness for purpose.

5. Works Cited

[1] Ullman, J., Eriksson, S.Å., Kåsin, J.I., Borgen, A., Reynolds, D., *Lab Testing of Suspension Seats for High-Speed Boats: Lab testing cannot determine seats' capacity to protect users from impact-induced injury*, RINA, High-Speed Vessels, 1-2 July, London, UK

[2] Coe, T.E., Finnemore, R.A., Smith, N.J., Protocol 1: Laboratory Testing of Shock Mitigating Seats, Naval Design Partnering Team, 2013

[3] Coe, T.E, Finnemore, R.A., Smith, N.J., On the Selection and Real-World Effectiveness of Shock Seats for High-Speed Craft, 51st UK Conference on Human Response to Vibration, 2016, Gosport, UK,

[4] ISO 19470 Working Group, Committee Draft 19470 (Unpublished)

[5] BSI Working Draft (Awaiting Publication)

[6] Marshall, J., Coats, T., Riley, M.R., The Rationale, Assumptions, and Criteria for Laboratory Test Requirements for Evaluating the Mechanical Shock Attenuation of Marine Shock Isolation Seats, Naval Surface Warfare Center, Carderock Division, August 2018

[7] UK Ministry of Defence, DEFSTAN 00-250: Human Factors for Designers of Systems, 2012

[8] Smith, N.J, Whole body vibration – seat testing metrics, UK Human Response to Vibration Conference, Gosport UK, 2016

[9] Riley, M., Haupt, K., Jacobson, D., A Deterministic Approach for Characterising Wave Impact Response Motions of a High-Speed Planing Hull, Naval Surface Warfare Center, Carderock Division, January 2012

[10] Riley, M.R., Coats, T.W., The Simulation of Wave Slam Impulses to Evaluate Shock Mitigation Seats for High-Speed Planing Craft, Naval Surface Warfare Center, Carderock Division, May 2013

[11] Riley, M.R., Coats T.W. and Murphy, H. Acceleration Response Mode Decomposition For Quantifying Wave Impact Load in High-Speed Planing Craft, Naval Surface Warfare Center, Carderock Division, April 2014

[12] Hirst, J., Dyne, S., Synthetic Vertical Acceleration Time histories for High-Speed Craft, The 8th International Conference on Surveillance, Search and Rescue Vessels, Royal Institute of Naval Architects, 2013

[13] EuroNCAP, Assessment Protocol - Overall Rating, June 2020, https://cdn.euroncap.com/media/58228/euro-ncap-assessment-protocol-overall-rating-v901.pdf

[14] Smith, N.J., Literature Survey on Injuries Caused by Non-Vertical Shocks, and the MADYMO Model, NDP71V440, Naval Design Partnering Team, 2018

[15] Riley, M.R., Murphy, H.P., Coats, T.W., A Consistent Wave Impact Load Model for Studying Structure, Equipment Ruggedness, Shock Isolation Seats, and Human Comfort in Small High-Speed Craft, Naval Surface Warfare Center, Carderock Division, 2016

[16] International Standards Organisation, ISO 2631-5 - Evaluation of human exposure to shock and vibration - Part 5: Method for evaluation of vibration containing multiple shocks, 2018.

[17] Mansfield, N.J., Impedance Methods (Apparent Mass, Driving Point Impedance, Mechanical Impedance, and Absorbed Power) for Assessment of the Biomechanical Response of the Seat Person to Whole Body Vibration, 2005, Industrial Health, 43, pp. 378-389

[18] Mansfield, N.J., Griffin, M.J., Non-Linearities in Apparent Mass and Transmissibility During Exposure to Whole-Body Vertical Vibration, Journal of Biomechanics, 33(8), 2000

[19] Patelli, G., Effects of vertical mechanical shocks and body posture on discomfort, Doctoral Thesis, University of Southampton, Faculty of Engineering and the Environment, 2016

[20] Griffin, M.J., The Handbook of Human Vibration, Academic London, 1990

[21] Dobbins, T., Myers, S., High-Speed Craft Motion Analysis - Impact Count Index (ICI), 43rd UK Conference on Human Responses to Vibration, 2008, Leicester, UK,

6. AUTHORS BIOGRAPHY

Phill Moxley MEng is a Whole Body Vibration analyst with the Naval Design Partnering Team, and as a Noise and Vibration Consultant Engineer at Frazer-Nash Consultancy. He has participated in ISO, BSI and NATO committees on high-speed craft shock mitigation standards, and undertook a four-year research posting on human responses to vibration at the Human Factors Research Unit at the University of Southampton. His work at the NDP has involved both laboratory drop testing and at-sea measurements on high-speed marine platforms.

Dr Tom Coe CEng MRINA, is a maritime whole body vibration specialist who has worked for a range of UK Government and non-government agencies in the UK and Europe as well as commercial customers on maritime WBV safety issues. He is the former technical lead for the UK MoD's Maritime Whole Body Vibration Project, and holds a PhD in human seat interaction from the University of Southampton and an MEng from the same institution and is the British Standards Institution WBV Panel Chair.

Richard Lawrence MEng MRINA

A Naval Architect and structural engineer with over nineteen years of experience in design, analysis and manufacture. Successfully delivered innovative solutions across a wide range of domains, from composite analysis of the Tamar Class RNLI lifeboat to fixed wing integration on Queen Elizabeth Class aircraft carrier.

REQUIREMENTS, INNOVATION AND PROCUREMENT: A 21st CENTURY UK PERSPECTIVE

J Louth, Subsea Craft Ltd, UK

SUMMARY

This paper explores the interactions between future defence requirements in a period of strategic uncertainty, the development of exquisite, technology-rich capabilities emerging from the private sector, and the potential changing face of government defence acquisition. It starts from the premise that the UK government's practices in relation to defence and security procurement, sometimes taking years in managerial assessments, are unwieldy and unfit for rapid technology insertion and the demands of the modern epoch. This inhibits, specifically, the ability of innovative design and emerging technologies to be deployed on the front line, potentially eroding UK defence capabilities. The manner in which this systemic, governmental inhibition is being challenged by niche companies in the private sector is explored by way of a review of the development of Subsea Craft Ltd, a small, UK business in its infancy. Taking an ethnographic approach, the work blends practices of horizon scanning, defence innovation and the commercialisation of ideas into the possibility of deployable defence capabilities.

1. INTRODUCTION: A PROBLEM POSED

From a Western perspective in the 21st Century, the practice of defence is exercised through complex interstate, military and public/private alliances and partnerships [1]. These are concerned with harnessing known and emerging technologies and innovative thinking to provide women and men with doctrines, hardware, software, knowledge and capabilities that are better than those of potential opponents, whether they be rival countries or non-state actors [2].

In this context, there are a number of reasons why advanced technology is particularly important to countries such as the United Kingdom (UK) and its emerging defence posture. For one, the UK cannot match the size of forces that countries such as Russia and China can muster. In addition, states that rely in military terms on mass human commitment and tactical ingenuity for success, in lieu of advanced technologies, normally have to endure sustained losses over an extensive period to avoid strategic defeat. In this regard, the lessons of the North Vietnamese and Vietcong in their victory over the United States resonate and, on the face of it, could never be replicated in the UK as deliberate government defence policy [3]. Moreover, in terms of alliances, there is value in UK forces fielding capabilities compatible with its main partner, the United States, especially in the setting of NATO. More generally, the Western, market-centric economic system has long had a comparative advantage in technological innovation and development over rival systems such as those associated with China. This comparative advantage is of course being ambitiously challenged and eroded by emerging powers. Given this, it is hard to conclude otherwise that British Governments will continue to place considerable emphasis on technology as a very prominent element of British military power.

Acquiring that advanced technology in an epoch of rapid technological innovation should prove relatively straight forward given the importance articulated above [4]. But the UK government struggles to innovate, secure and promote the disruptive, niche technologies necessary to secure or maintain UK military advantage. Rather, innovation and technological development resides often in the private sector with government seemingly systemically ignorant of what is being undertaken and achieved. Harnessing notions of governmental need to technological advances offered by private sector initiative and capital is critical but poorly understood.

Taking an ethnographic approach to consider these issues [5], the paper considers Subsea Craft Ltd, a small, niche UK business concerned with the design, generation and delivery of battle-changing maritime technology to Western governments and allies [6]. This is the most obvious methodology to be utilised given that the author is a non-executive director of the business. He has observed first-hand the intent of the executives in bringing-forward niche, technology-rich capabilities to fill the armouries of the West, and the challenges encountered. The paper begins with the argument that Subsea Craft chooses not to align to conventional Defence procurement practices as, to do so, would prevent the very development of the technology-rich capability necessary for UK success. The impact of accepting, deliberately, such a massive commercial risk, is acknowledged. Thereafter, the paper explores the business's approach to technology development, specifically focussing on the assessed needs of the future operating environment. The paper follows this with conclusions around this manner of capability generation and the implications for government procurement and the stewardship of defence research and development.

2. THE PRACTICE OF CONVENTIONAL DEFENCE PROCUREMENT IN THE UK

Conventional defence procurement in the UK is a formulaic, bureaucratic affair, especially witnessed in relation to the so-called "smart acquisition" reforms associated with the Blair governments. The stated objectives of smart acquisition were clear [7]:

- To deliver projects within the performance, time and cost parameters approved at the time the major investment decision was taken.
- To acquire military capability progressively, at lower risk, and with optimisation of trade-offs between military effectiveness, time and whole life costs.

This alphabet soup of managerial expositions also contained the hope that smart acquisition initiatives would cut the time for key, new technologies to be introduced into the frontline, especially where needed to secure military advantage or UK industrial competitiveness [8].

The ambition was swamped by harsh reality. Defence projects could take decades to mature with the National Audit Office continually criticising the Ministry of Defence for project delays, overspends and management incompetence. Whilst some reforms in practice were seen to add value, smart acquisition was often dumb in its delivery, with new and disruptive technologies, at best, randomly finding their way to the military operator [9]. This, of course, during a time when the UK and Western allies had deployed their forces on major, multi-year campaigns in Iraq, Afghanistan and Libya.

The problems with defence procurement now, in the second decade of the 21st Century, could probably be characterised in the following way. Defence acquisition reform in the past 30 years has been focussed on the avoidance of waste and the generation of evermore efficiencies and economies. This yielded a highly managerial, programmatic approach to the generation of defence capabilities, linking inputs to outputs in a manner that excluded notions of strategic effectiveness or critical national research and development investments. Absent has been a sense of urgency, in relation to the harnessing of tomorrow's technologies, or a realisation of the need to generate a culture that had technological innovation, and matching investment, at its heart.

Ignoring established defence procurement processes, commanders often utilised what were known as 'Urgent Operational Requirements' to purchase equipment, either on the eve or during a major operation. Investing in emerging technologies for tomorrow's requirements often went by the way. This ignored the strategic imperative that, in the light of the assertiveness and growing capabilities of potential adversaries, such as Russia and China, the need for innovation at component, sub-system and system levels should have been, with hindsight, the major defence objective.

Instead, the agencies of smart acquisition favoured centring on the notion of a "user requirement" and a "systems' requirement" to prompt industry to bid, typically in competition, for contracts to be awarded. If a particular niche technology was not aligned to a successful bid team then, often, that technology could be lost to defence decision makers [10]. For the private sector therefore, as suppliers to the MoD, business relied on government stating a requirement and then running a commercial competition under a 'winner takes all' mind-set. Many small businesses and technologists new to the defence market found this a major disincentive to stay in the sector [11]. Not surprisingly, innovation wandered to more conducive watering holes, or withered away [12].

3. ADVANCING CHALLENGER TECHNOLOGIES

The founders, financiers and executives of Subsea Craft Ltd reasoned that the dynamics of the conventional UK defence procurement system would inhibit the ability to bring forward niche technologies. Consequently, the business decided to self-fund development of the VICTA class (which is discussed below) to Technology Readiness Level 5, accepting overtly the commercial risk held as a consequence of this decision. The reason was two-fold: the business believed that this capability is critical to the defence posture of Western forces in the future operating environment. Second, that without the commencement of development of VICTA outwith an overt government requirement, the capability would not be available to the frontline in the mid-2020s, when it would probably be needed. Without being too dramatic, the commercial risk could also be characterised as the patriot's risk, where private finance believes that it has horizon-scanned effectively, and with knowledge of emerging technologies, to predict capability requirements for defence in advance of governments. If this is correct, an element of defence innovation in the West rests on the fragility of the willingness of private finance and technology researchers to back their own hunches.

It is this essence of a business believing that it understands the future operating environment and the technologies needed that coagulate to form tomorrow's emerging defence capabilities. In the case of Subsea Craft, it is worth exploring these sensed and reasoned conditions.

3.1. THE FUTURE OPERATING ENVIIRONMENT

Until at least 2050 or so, the operating environment in defence is likely to be characterized by great power competition, especially between China and the United States. Other states and nations could well be impacted by this and middle-ranking powers across NATO, the Middle East and the Pacific may well become embroiled. Tensions between the US, Europe and Russia will also continue with new cyber and information operations' threats probably emerging from Moscow. In parallel, this era will see multiple proxy wars and indirect engagements across the spectrum of conflict.

This trend will be enabled by the mass availability of sophisticated intelligence, surveillance, reconnaissance and weapon systems, along with their extended ranges. The proliferation of these systems, and their availability to entry level actors, will allow proxies and states at every level to create bubbles of control around their critical interests – an extension and evolution of the concept once known as Anti Access and Area Denial [13]. Consequently, there will be a need to counter capabilities that prevent Western forces from operating in these bubbles, or to provide assets that can operate stealthily, in swarms or alone, hiding within the 'bubbles' being penetrated.

Allied to this is the realisation that asymmetric, highly flexible assets will need to be at the heart of military strategy in the new operating environment. Whilst there is, of course, a place for large conventional warships, for example, such capital systems are vulnerable to a modestly capable adversary and should have less reliance placed upon them [14].

The confluence of these major evolutions in the future operating environment will require militaries to provide Western politicians with a different set of potential responses to aggression. Instead of seeking, as a first option, an overt military engagement, the future force will be working on and in the seams between the traditional defence domains of the land, sea and air. For example, Hypersonic weapons have been so successful principally because these weapons fly in the seam between endo- and exo-atmospheres (air and space). If working the seams of each conflict becomes a prerequisite for success, achieving penetration from the surface/subsurface seam offers the greatest opportunities for influence, action and the generation of politically positive outcomes.

Almost every dilemma politicians and militaries will face out to 2050 will require a means to penetrate and burst bubbles of control surrounding the geographic area of the disputed region – in the air, in space, on and below the sea, as well as on the land. Doing this from the sea is, currently, an under exploited opportunity.

Historically, this seam has been exploited by navies, marines and special forces through the use of swimmer delivery vehicles. However, these platforms – well proven though they are – have some significant restrictions in how they can be used, relating to speed, range, payload, and endurance. Rather, a new platform is required that overcomes the previous operating envelope of these assets. This is not simply a large hull submarine capability, but rather lends itself towards a cheaper off-the-shelf design that can be operated alone or in multiples, as necessary.

The asset must enable an extended over the radar horizon deployment, fast insertion, with a low radar profile and minimal electro-magnetic signature, a crew payload of at least 6 operators (or a NATO one ton pallet), the option to submerge at key points of radar coverage, the ability to loiter or remain on station, and to covertly self-deploy drones, buoys and bodies into the adversary's bubbles of control (and depart without detection). Such systems may well work alone (as with current platforms), but increasingly will work in teams or swarms to overwhelm the enemy's kill chain system.

Ideally such assets will need to be forward deployed and maintained in several locations globally in order to meet the challenges of multiple dilemmas posed by adversaries. They will need to be air transportable by commercial and military aircraft, or by standard ISO containers, and should exceed vertical airlift and deployment restrictions associated with current military helicopters.

Operation will require a simple Human-Machine-Interface, ideally designed by military operators who will understand the realities of system operation under considerable stresses. Moreover, options around maintenance and ownership will form a key part of the system solution for such missions. Given the financial models many militaries operate under, opportunities exist for a partnership with industry to minimise purchase price, and to contract for availability, or to have the craft commercially maintained and forward deployed by a trusted commercial partner [15].

Subsea Craft Ltd foresaw that military challenges of the next 20-30 years will require a capability in the West to penetrate and operate successfully within an adversary's area of control. This will be critical to secure or regain the political and military initiatives from competitors below the threshold of armed conflict, typically exercised through the traditional tools of sabotage, subterfuge, and deniable action. In times of armed conflict itself, such an asset will allow operations in contested areas with minimal risk to large, capital assets and the crews who serve in them.

The business was formed on the insight that there was a need, currently unstated by governments, for an asset that grows and builds military response options for political leaders, with overt, covert and deniable access to contested areas at speed and range greater than those previously imagined. Such a platform would enable new methods of horizontal escalation, de-escalation and coercion that have previously been bounded by platform restrictions in the seam between the sea and subsea domains.

Bringing this vision to the frontline is based on an informed bet that the horizon viewed by executives and investors un Subsea Craft Ltd is correctly perceived and that Western governments will seek the VICTA as a necessity to operating within the environment described. Innovation in this manner, beyond government, in an epoch of rapid technological change, has profound implications of course for ministerial procurement systems around the world. For the procurement processes derived from smart acquisition in the UK would not bring forward a VICTA: a statement that should concern everyone interested in Western defence.

3.2. THE VICTA CLASS

To address this sense of need for Western forces, Subsea Craft Ltd bet its multi-million-pound research and development budget, sourced from private investment, on doing three things:

- Building a physical craft, the VICTA class, as the centre-piece of fresh, critical capability.
- Developing a virtual, digital fingerprint of all enabling technologies, components, design, manufacturing and maintenance schedules and operating doctrines/protocols to enable training, development, maintenance, rectification, mission planning and operational simulation. This is both for the cost benefits of ownership but also to enable network applications and solutions with VICTA as a node of a broader defence environment and effort.
- Building a team at Subsea Craft Ltd that will be with the VICTA journey for the long term, enabling the provision of real expertise in operations, manufacturing options and maintenance for governments and military clients.

At its most orthodox, VICTA is a surface/sub-surface maritime insertion and interdiction craft. Characteristically, it possesses an independent surface long-range insertion capability of 250 nautical miles at a speed of 30 knots depending upon the prevalent sea state. On arrival at a surface to sub-surface transition point, VICTA can convert to a sub-surface asset within two minutes. The craft is configured optimally to cruise sub-surface at six knots for a range of 25 nautical miles. Whilst the normal operating depth is six to ten metres, the craft is pressure tested to a depth of 50 metres. Consequently, a maximum operating depth can exceed significantly the normal operating depth when a flexible response to adversaries is demanded or the mission outline so requires.

Therefore, the orthodox use of VICTA is the clandestine insertion and recovery of special forces or maritime operators into a target area, utilising personal breathing apparatus. In this regard, the command module has capacity for one pilot and one co-pilot, both of whom can operate the craft independently, whilst the payload compartment houses up to six special forces operators.

VICTA possess clearly a significant utility below the threshold of actual warfare. As stated above, national decision makers continue to articulate a need to maintain or regain a political or military initiative in relation to competitors through traditional state tools of intelligence, sabotage, subterfuge and other targeted interventions. VICTA offers to leaders ever-evolving options of military responses, with covert and deniable access to rivals at a speed, range and level of control that is properly disruptive of what was thought possible to-date.

Personnel under transit within the craft utilise shock mitigation seats and adjust buoyancy to maximise comfort. VICTA has also the capability to transport dense negatively buoyant equipment or payloads maximising its utility and flexibility. The craft, therefore, provides a portal to new avenues of state-to-state escalation, de-escalation and considered, tolerable risk-taking that have been bounded previously by restrictions associated with conventional platforms and large maritime capital assets.

Predictions relating to the future operating environment revolve, principally, around notions of contested space in key areas of the world whether this be, for example, the Baltic or the South China seas. The contest to these key areas will not be limited to operations by large military powers but also by the interests of small and medium-sized powers implicated by geographical proximity, treaty obligations or raw self-interest. At every phase of conflict, the ability to engage in a theatre at a time and place of choice is an ambition that is now challenged seriously by the proliferation and democratisation of sea, land and aerospace denial systems, as has been discussed. With access to critical theatres of operation now a widespread problem for militaries, finding assets that can "burst the bubble" of adversarial control, or to covertly monitor enemy movements and deployments within such bubbles, becomes a critical enabler for the execution of all military options. VICTA provides such a solution for "bursting the bubble" whilst, concurrently, offering the asset from which multiple mission options can be conceptualised and launched [16].

It is reasonable to assume that political leaders in the Western democracies do not wish to become embroiled in military operations at whatever threshold of conflict. Moreover, it does not seem overly idealistic to state that a world without interstate rivalries, strategic competition and the potential for armed conflicts is a preferable condition for humanity. In this regard, it is better to deter and dis-incentivise potential aggressors. VICTA's flexibility and broad range of utility, the covert nature of its operations, its deniability and affordability earns it the right to be considered as an asset capable – and necessary – to be included overtly within discussions of modern deterrence.

4. CONCLUSIONS

Traditionally, the manner in which government procures defence capabilities has been a bureaucratic endeavour centre on a statement of military requirement and focused on themes of efficiency, economy and value for money. These are important, of course, but it is an agenda that hinders rather than harnesses proper innovation and disruptor-technology insertion.

One contention of this paper is that there is a correlation between urgency and innovation that has escaped decision makers until recently. For example, in the UK, the Defence Solution Centre, an initiative jointly funded by government and industry, has been developed to offer niche technologies to the armed forces, champion exports to friendly states, and to focus on collaborations to enable the fielding of innovative solutions at the earliest point possible [17]. This is a significant change. VICTA is a response, in part, to this spirit of acquisition which champions speed and operational and technological innovation over conventional bureaucracy.

A defence procurement system to best fit this spirit of innovation and risk taking should possess the following characteristics:

- The recognition that continual innovation and rapid, advanced technological insertion into defence systems is critical to the West's military advantage. This is more significant, strategically, than good commercial supplier management, the husbandry of resources or the continual, ever-present generation of efficiencies. Witnessing the development of the Defence Solution Centre in the UK and the Defense Innovation Board in the US has signalled governments' increased understanding of this critical point. Nonetheless, a capability such as VICTA was still, of necessity, developed by private sector insight and risk taking unreferenced or supported by government.
- An understanding that major systems need to be conceptualised and designed overtly as open systems, enabling the development and deployment of new technologies. VICTA needs to integrate into an existing order of battle, operating as both a nodular and standalone asset, being designed specifically to this standard.
- A commitment that the financing of capability-specific technology demonstrators should be shared between government and industry. As part of its proof of concept, at an agreed level of technological maturity, VICTA should be able to access the MoD's Innovation Fund, with a frontline command (or proxy) offering range support and open water facilities within the trial phase of the technology demonstrator. This will ensure that private capital and high-net worth individuals continue to take calculated risks on the development of technologies critical to defence.
- The vision that government's approach to risk management within defence acquisition needs to evolve. Disruptor businesses with niche technologies and systems to offer, but with a short trading history, evolving organisational structures and private finance, should be valued by decision makers as much as large, established international defence businesses, wherever they happen to be headquartered.

This recognition, understanding, commitment and corresponding vision would do much to ensure that the West will be able to field the capabilities required in the years ahead, especially if they are to be developed on-shore in the UK. In itself, beyond defence, this might be critical if the UK government is to meet the ambitions of its prosperity agenda and ensure that knowledge-rich jobs are nurtured and developed post-COVID 19 and as the country sets course for new ventures, having left the European Union.

5. ACKNOWLEDGEMENTS

The author would like to acknowledge the decade of collaborative work with Professor Trevor Taylor, and recent joint research with Professor Peter Roberts, both of RUSI. These collaborations have informed this paper. Gratitude is also due to the executive team of Subsea Craft Ltd for allowing the author to access and share company strategies.

6. REFERENCES

1. GANSLER, J.S., 'Democracy's Arsenal', *MIT Press*, 2011.

2. HEIDENKAMP, H., LOUTH, J., and TAYLOR, T., 'The Defence Industrial Triptych', *RUSI Whitehall Paper 81*, 2013.

3. HASTINGS, M., 'Vietnam: An Epic Tragedy 1945-1975', *William Collins*, 2018

4. ATKINSON, S., and MOFFAT, J., 'The Agile Organization', *The US Command and Control Research Group*, 2005

5. SAUNDERS, M., LEWIS, P., and THORNHILL, A., 'Research Methods', *Prentice Hall*, 2000.

6. www.subseacraft.com accessed on 30th December 2020.

7. MINISTRY Of DEFENCE, 'Smart Acquisition Handbook', *Acquisition Management System*, 2002.

8. *Ibid.*

9. CLARKE, M. (Ed), 'The Afghan Papers', *RUSI Whitehall Paper 77*, 2011.

10. AUGUSTINE, N., 'Augustine's Laws and Major System Development Programs', *Defense Acquisition Research Journal*, 2015.

11. LOUTH, J., and TAYLOR, T., 'THE US Third Offset Strategy: Hegemony and Dependency in the Twenty-First Century', *RUSI Journal*, 2016.

12. LOUTH, J., TAYLOR, T., and TYLER, A., 'Defence Innovation and the UK: Responding to the Risks Identified by the US Third Offset Strategy', *RUSI Occasional Paper*, 2017.

13. LOUTH, J., and ROBERTS, P., 'The Victa Vision – Next-Generation Capability: Adaptable, Affordable and Resilient,' *Subsea Craft*, 2020

14. MARTINAGE, R., 'Towards a New Offset Strategy', *US Centre for Strategic and Budgetary Assessments*, 2014.

15. LOUTH, J., and ROBERTS, P., 'The Victa Vision – Next-Generation Capability: Adaptable, Affordable and Resilient,' *Subsea Craft*, 2020

16. *Ibid.*

17. www.ukdsc.org accessed on 31st December 2020.

7. AUTHOR'S BIOGRAPHY

Professor John Louth is the senior non-executive director for Subsea Craft Ltd and a specialist adviser to the House of Commons Defence Select Committee. His specialism is the defence enterprise, defence acquisition and technology management. He was a research director at the Royal United Services Institute in London, has advised governments around the world on defence capability management and taught and supervised research at a number of universities. Until this year, Professor Louth was the UK representative on the research board in the EU assessing European defence capabilities. He served in the British military between 1987 and 2004.

RNLI SEVERN CLASS LIFEBOAT LIFE EXTENSION PROGRAMME

N Chaplin, N Fenwick, C Smith, I Sands, T Coe, F Prini* and **H J Phillips,** Royal National Lifeboat Institution, UK

*Current affiliation: QinetiQ, Haslar, UK

SUMMARY

The Severn class lifeboat was introduced into RNLI service in 1995 with a nominal design life of 25 years and was the first RNLI all-weather lifeboat to be built from advanced composite materials. 25 years later the fundamental operational requirements have not changed, which has presented the RNLI with an opportunity to consider a life extension of the boats for a further 25 years.

The paper will show how the RNLI has assessed the structural integrity of the hulls to validate life extension and how re-design of the on-board systems aims to deliver a lifesaving asset suitable to face new challenges and expectations of the next 25 years in a cost effective and sustainable manner, whilst acknowledging practical limitations and compromises.

NOMENCLATURE

Δ	Boat displacement (kg)
λ	Wave length (m)
LCG	Centre of gravity longitudinal position (m)
LOA	Overall boat length (m)

1. INTRODUCTION

At the time of writing, the Royal National Lifeboat Institution (RNLI) are undertaking a project to life-extend the largest of its lifeboat classes, the Severn class. This paper sets out the rationale and the considerations made so far, in bringing this proven life-saving asset up to date and ready for another 25 years of service in saving lives at sea.

The paper will show how the project aligns with the RNLI's future aims and how the life-extension option represents good value compared to the typical replacement of lifeboat tonnage at the end of life.

Whilst work has begun on the physical updating of the first Severn class boat in the Severn Life Extension Programme (SLEP), the full design work is not finally complete and tests and trials of the first boat will be required to validate and prove the design and refit work.

2. WHY ARE THE RNLI LIFE-EXTENDING THE SEVERN?

2.1 OPERATIONAL BACKGROUND

The RNLI voluntarily operates a fleet of all-weather lifeboats (ALBs) and inshore lifeboats (ILBs) around the coast of the United Kingdom (UK) and Republic of Ireland (RoI) to conduct search and rescue (SAR) activities as requested by the co-ordinating authorities in those countries. The RNLI provides a seasonal beach lifeguarding service in conjunction with local authorities at selected locations around the UK.

To fulfil this function, the RNLI requires a range of lifesaving assets tailored to the specific operational and geographical requirements, including a fleet of 150 ALBs strategically located around the coast. Fig 1 below provides an overview of the locations of RNLI operational stations around the UK and RoI.

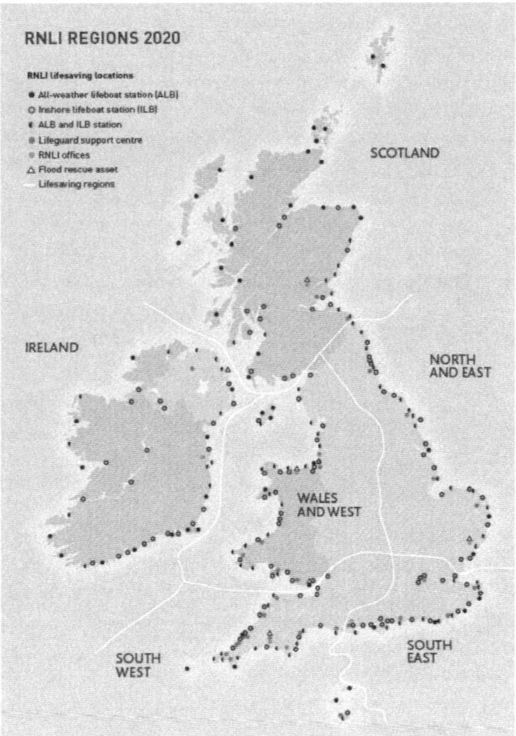

Fig 1. RNLI Operational Stations

The operational requirement for all-weather lifeboats (ALBs) was last increased in the mid-1990s from a maximum speed of 18 knots in calm weather and to operate out to 50 nautical miles (Nm) offshore, to a maximum speed of 25 knots in calm weather and to operate out to 100Nm offshore in all weathers.

To achieve the increased range and speed, the RNLI undertook a development programme to replace 18 knot tonnage at the end of its service life with new boats capable of meeting the requirements from an afloat berth, slipway or carriage launch. The outcome has been the creation of the Severn, Trent, Tamar and Shannon class ALBs. Each class can achieve the 25 knot, 100Nm operational requirement, and each is optimised for the specific challenges of launch, recovery and berthing faced around the coast of the UK and RoI.

3. WHAT IS THE RNLI AIMING TO ACHIEVE?

3.1 RNLI STRATEGIC AIMS

The RNLI undertakes periodic reviews of its business plan and longer-term strategies, the latest of these activities delivered a new plan titled *Our Watch* in early 2020. This document sets out the RNLI's aims to 2024 and covers some longer-term aspirations up to 2050, to help achieve the RNLI's Vision: to save every one.
6 outcomes are defined in *Our Watch*:

- We keep our people safe.
- We focus on what will save most lives in or near the water.
- Working with partners, we influence improved safety on and beside the water.
- Our people love being part of the RNLI.
- We raise the funds we need and spend donor's money wisely.
- Our future is sustainable and secure.

Whilst the SLEP project was initiated before the publication of *Our Watch*, the project seeks to support the delivery of the outcomes too. More specific details on some topics are provided later in this paper.

<u>We keep our people safe</u>

The RNLI relies on its volunteer crews and lifeguards to save lives at sea and has a duty to ensure their safety and the safety of those who support them. Lifesaving in an unpredictable environment is inherently hazardous, and even with the best boats, training and kit, accidents can happen. The following are examples of safety improvements being addressed as part of SLEP.

- Whole Body Vibration (WBV) and ergonomics.
- Integrated electronics to provide crew with more monitoring and control of the boat whilst seated.
- Replacement of currently manually operated deck systems such as emergency steering and man-overboard recovery arrangements.

<u>We focus on what will save most lives in or near the water</u>

The RNLI are a search and rescue service and remain focused on being there to save people in need. Every year over 40,000 people are assisted in, on and near the water. The RNLI know that most lives are lost close to shore, however the need remains to provide strategic lifeboat cover to attend to less frequent but demanding maritime distress incidents further offshore.

Although all RNLI ALB classes can achieve the same operational requirement, the Severn offers some advantages over and above the capability of the current production ALB, the Shannon class. Specifically, the Severn is larger and more powerful and therefore has increased capabilities such as a larger survivor handling capacity, more space for casualties and a greater towing bollard pull. Additionally, the greater size and displacement offer a better platform for crew undertaking long endurance shouts such as prolonged searches or towing operations from further offshore.

To save lives, it is necessary to identify, locate and assist casualties. To this end, the SLEP project will implement improvement in several aspects, including:

- Searching – New searchlights and sensors.
- Daughter boat – New daughter boat and launch and recovery crane.
- Navigation and communications upgrades – Obsolete stand-alone navigational packages will be replaced with RNLI systems.
- Connectivity – The upgrade will deliver data and information sharing capabilities.

<u>Working with partners, we influence improved safety on and beside the water</u>

RNLI research on the Drowning Chain shows that, even if there was a lifeguard and lifeboat on every beach and every stretch of water, the RNLI would not be able to save everyone through rescue alone. To achieve the Vision the RNLI must work hard to stop people getting into trouble in the first place and help them make the right choices to reduce their risks.

Whilst a rescue asset can't intervene directly in this preventative work, it can play a role in collecting and sharing valuable data which can contribute to large data sets for analysis to help find previously unknown information and patterns which will help inform future prevention work. The network of integrated sensors and systems, along with the connectivity capability to be fitted to SLEP will support more data collection than the RNLI has previously achieved.

Our people love being part of the RNLI.

The RNLI is reliant on volunteers selflessly giving time to serve as lifeboat crew and in other supporting roles to deliver the lifesaving effect. In recognising that, it is vital that the SLEP design pays attention to the needs of the volunteer and to how they wish to interact with the vessel now and into the future.

Patterns of volunteering are changing, turnover and throughput is likely to increase in future as volunteering patterns with charities changes. The maritime experience volunteers have is typically reducing and as such it is essential to consider the levels of knowledge, training time and availability of volunteers to carry out SAR missions safely and effectively.

The implementation of an improved wheelhouse layout to accommodate a greater anthropometric range, improved shock mitigating seating and the implementation of the in-house electronics system (compatible with other classes of lifeboat) will help to deliver a more inclusive, safe and simpler operational environment to carry out the lifesaving service.

We raise the funds we need and spend donor's money wisely.

It cost £163.5M to run the RNLI in 2018 (over £447,000 (€504,000) a day). Funds are voluntarily donated to the RNLI and investment in assets and infrastructure form a major part of the utilisation of those funds. The possibility of delivering an improved rescue asset which successfully addresses the aims of the organisation for around half the cost of a new asset is an attractive proposition.

RNLI new assets are generally funded by donations from supporters and can often be funded either wholly or significantly by single donations or legacies, however the RNLI has not considered the funding mechanism for a life-extended asset before.

When the original donations were made for the Severns, it was on the basis that the boats would have a service life of 25 years, after which they would be replaced with new assets. Now that the RNLI can confirm that the operational requirement has not changed and that the boats have the capability to carry on in service, the life-extension funding options have been reviewed and the decision made that, as the costs of life-extending are still significant, it was appropriate to seek major donations similar to fundraising for new boats.

Our future is sustainable and secure.

In 2024, the RNLI will be celebrating its 200th anniversary and it must ensure that it is set up to continue to save lives for the next 200 years. The RNLI is committed to considering sustainability across all functions, economically, environmentally, and socially. The environmental ambition to eliminate or reduce the negative environmental impacts has been a central consideration in SLEP.

Following the environmental 3R's (reduce, reuse, recycle) approach, the re-use of the Severn hulls reduces the consumption of construction materials. Life-extension also delays the disposal of the composite structure until a time when technologies and processes will probably be available to dispose of it in a less impactful way. By not building another new asset now, the replacement is also potentially moved into the future, where again it is believed that more sustainable construction materials and environmentally friendly propulsion solutions will be available.

The life extending process will require the removal and disposal of various parts of the existing boats, which will create waste. The RNLI are reviewing every item removed and assessing the impact of disposal with an aim to minimise the volume of waste. Where possible components are being refurbished and reused or recycled.

The subject of the challenges regarding providing a more sustainable propulsion solution are covered in later sections of this paper.

4 OPTIONS FOR SEVERN REPLACEMENT

4.1 NEW BUIILD

In planning how to manage the end of life of the Severn class, the RNLI considered options including building new lifeboats to replace them. The factors considered included:

- What are the changes in operational requirements?
- What new technologies are available to improve on the original design and deliver the strategic aims?
- What would the through life cost of a new build be?

As there were no changes in operational requirements for the replacement of the Severns, no fundamental enhancements were necessary to continue to achieve the lifesaving goals.

For new technologies, the improvements sought were in the fields of alternative propulsion, improved structural materials and the management of WBV. Initial investigations into alternative propulsion did not identify any suitable proposals. Equally, no major developments in structural materials appeared to offer significant advantages over the current glass and epoxy materials used in RNLI lifeboat building. There have been however significant developments in the management of WBV particularly in shock mitigating seating and other ergonomic improvements, but these alone did not necessarily warrant a new build option.

Through life costs. The two major cost elements of a vessel are typically the initial design and build, followed by the operational and support costs of the lifeboat. Given that the initial analysis of a new build option did not identify any realistic significant changes in structure nor propulsion, it is likely that the operational and support costs for a new build would be similar to those of a life-extended version of the Severn. The major difference in overall costs would therefore be in the design and build.

A brief analysis of lifeboat build costs in the RNLI and from previous yard contracts, corrected to 2020 prices using UK inflation rates, shows that the likely cost of a new build ALB of the size of a Severn would be around £3M. If the same fleet size were to be replaced, then 44 boats would require an investment of around £132M.

4.2 LIFE EXTENSION OPTION

The Severn and Trent classes were designed concurrently in the early 1990s at a time when experience in the use of advanced composites was limited in the maritime field, and structural rules for these materials were still in development. The work was underpinned by research by [1] [2] [3], materials testing and in build inspections using both external auditors and RNLI engineers. The final design solution, which the RNLI believed would provide a robust lightweight structure capable of achieving the increased operational requirements, was implemented into production [4].

The option to life-extend the Severn class is only viable if the structural arrangement can be validated to be suitable for another 25 years of service. Accordingly, the RNLI began a programme of research into lifeboat structural assessment culminating in the publishing of a thesis on the residual life assessment of composite structures with application to all-weather lifeboats, Roberton [5].

Since introduction into service, both the Severn and Trent lifeboat classes have proven to be successful and robust rescue assets. The Severn has been involved in many dramatic and notable missions from the rescue of the crew from the Green Lily off Shetland in 1997, to working close inshore in the surf off Newhaven to rescue a person in the water. Additionally, the performance of the Portrush Severn when grounded during an operational call to Rathlin Island in 2008, proved the robust and trustworthy nature of the Severn design and build in very challenging circumstances [6].

Fig 2. Images of Severn class lifeboats in active service

4.2 (a) Known Risks and Issues

No operational asset is entirely free from risk or design compromise, and as such the RNLI undertake safety assessments of the designs of operational assets to identify, eliminate or treat risks and create a residual level of operational risk that is as low as reasonably practicable (ALARP).

The initial design of the Severn was undertaken in the early 1990s when formal safety assessment was less mature than in most organisations now. Operational experience and improved understanding of risk and the creation of new risk management tools has enabled a more complete and thorough assessment of operational risk.

The SLEP project has utilised a risk management tool developed by the RNLI with an international SAR partner, the KNRM, who carry out maritime search and rescue in the Netherlands. The tool developed is named SHARK-DAT and enables organisations to assess regulatory, standards requirements and risk management in a coherent manner to identify, assess and manage risks. The background and development of SHARK-DAT is covered in [7]

The output of the SHARK risk review highlighted some areas as requiring improvement, to improve on levels of safety, effectiveness and sustainability:

- Crew safety – Whole Body Vibration, ergonomics and anthropometrics.

- Structural capability – Ageing structure, unknown long-term material properties.
- Electrical safety – Ageing wiring.
- Daughter boat crane – Control system improvement.
- Mechanical systems – Obsolescence.
- Propulsion – Emissions targets.

The SLEP design team have identified improvement options to reduce the overall risk level associated with the operation of the Severn class.

4.3 OPERATIONAL LOADS AND STRUCTURAL ADEQUACY

Following the study from Roberton [5] on the residual operational life of the vessel's structure, further work was required to assess the structure's adequacy to sustain the seakeeping loads experienced over the whole extended operational life of the Severn. This was achieved by predicting with confidence the loads sustained by the structure during operation. Due to the challenges in predicting hydrodynamic loads at higher (semi-planing and planing) speeds and in extreme sea conditions, a collaborative research was undertaken with Newcastle University and Lloyd's Register to investigate the loads experienced during operation and the consequential structural response.

A systematic approach based on computational and experimental methods was developed to enable assessment of loads and response [8]. Seakeeping simulations, model-scale tests and full-scale trials were conducted to calculate static, slowly varying and transient loads, which were then superimposed on a structural model of the craft. This consisted of a global finite element model inclusive of all the main structural components and relative material properties (Figure 3). The response of the structure was calculated through finite element analysis and assessed to produce a set of structural response curves to enhance the insight into the operational envelope of the Severn [9]. The structural model of the craft and the experimental data were further exploited to support the design of the new Severn.

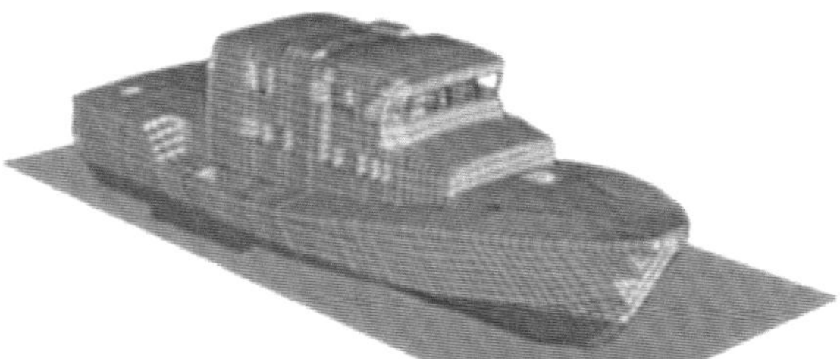

Figure 3. Global structural model of the Severn Class (from [9]).

The seakeeping simulation model, validated against the model-scale tests [10] and the full-scale sea trials [11], was used to advise, through a parametric study, on the effect of 'weights and centres' on the seakeeping performance of the Severn. With the limitations of the model accounted for (see [10, 11]), simulations were conducted by systematically varying the displacement (Δ) and the longitudinal position of the centre of gravity (LCG). The impact on the vessel's motions and loads was assessed through inspection of the response amplitude operators (RAOs) for heave, pitch, vertical bending moment and vertical shear force. An example of RAO plot displaying the effect of a shift in LCG on the pitch motions is shown in Figure 4, whilst a qualitative summary of the results is shown in Table 1.

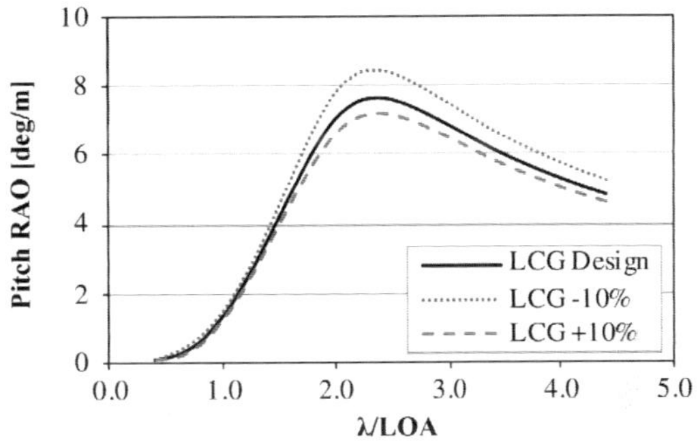

Figure 4. Pitch RAOs in head seas at 15kts for a 10% shift in LCG with respect to the original design value.

			Δ - 10%	Δ +10%	LCG -10%	LCG +10%
MOTIONS	Heave	Low speed (5-10kts)	~	~	↓	↑
		High speed (15-25kts)	~	~	↑	~
	Pitch	Low speed (5-10kts)	↓	↑	↑	↓
		High speed (15-25kts)	~	~	↑↑	↓↓
LOADS	Vertical Bending Moment	Low speed (5-10kts)	↓	↑	~	~
		High speed (15-25kts)	↓↓	↑↑	↓	~
	Vertical Shear Force	Low speed (5-10kts)	↓	↑	↓↓	↑
		High speed (15-25kts)	↓↓↓	↑↑↑	↓↓↓	↑↑↑

Table 1. Effect of displacement and centre of gravity position on motion and load responses.

The finite element model of the Severn was exploited to identify structural 'hot spots' (areas of high stress or strain) and inform required structural alteration to reduce the stresses (or strains) to lower levels. This task consisted of structural analyses at a global and a local level. The coarse-mesh global model of the Severn, which was suitable for applying load envelopes and assessing the global structural response, was first employed to identify potential hot spots (Figure 5). Fine-mesh local models of the hot-spot areas were then constructed to assess the actual magnitude of the stress (or strain) concentrations. An example of local finite element analysis is shown in Figure 6 for the hull bottom structure in way of the survivors' space.

The structural response was assessed according to a set of failure criteria suitable for composite laminates. In Figure 5, which shows the Tsai-Wu failure criterion, values that are greater than or equal to 1.0 (grey colour) imply first ply failure. In assessing the results, consideration was also given to the following aspects: load redistribution within the laminate following loss of the first ply's carrying capacity; effect of boundary conditions, which typically give rise to fictitious stress concentrations in way of the restraints; and presence of manufacturing features that, although not modelled, would contribute to the load transfer within the structure. With these aspects accounted for, the global and local structural models of the Severn were used to inform the level of structural adequacy of the Severn for operation over the next 25 years and support, where required, the design of structural modifications.

For this task, the loads applied to the hull were static pressures equivalent to slamming pressures with a return period of 50 years, which correspond the extended operational life of the Severn. These pressures had been calculated by converting strain measurements, taken during the seakeeping trials, to equivalent static pressures using local finite element models of the instrumented structure. Pressure values with a return period of 50 years were then estimated by means of Gumbel probability distributions and correction factors to account for the actual usage of the fleet over the 50 years.

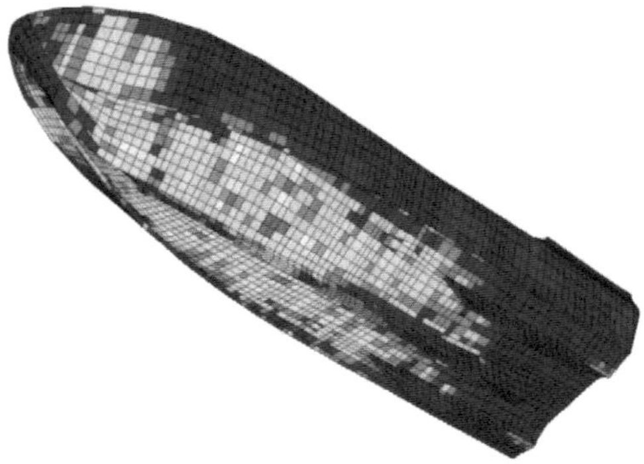

Figure 5. Global structural response analysis showing potential high stresses in the hull bottom structure in way of the survivors' space.

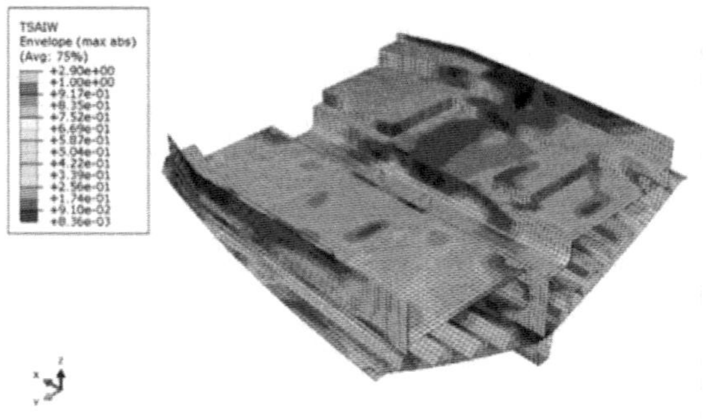

Figure 6. Local structural response analysis of the hull bottom structure in way of the survivors' space.

The calculated loads, which being based on a limited set of dedicated seakeeping trials were considered as approximations, were confirmed through comparison with the original design loads adopted for the Severn. Analysis of the structure's behaviour under the 50-year pressures thus provided sufficient confidence to proceed with the life extension of the original structure by embedding, as required, local structural alterations. The results of this analysis will also be factored into the structural inspections of each Severn lifeboat prior to completion of the upgrade work, as described in section 5.4.

4.4 OPTIONS ASSESSMENT

4.4 (a) Criteria

In establishing how the decision to new build / life extension would be made, the team based the work on standard Asset Management criteria of cost, risk and performance, supplemented with considerations of how each of the *Our Watch* outcomes would be supported.

Cost

The total through life costs of the fleet of assets must be considered, as well as cash flow affordability at the time of construction. Because the RNLI builds, refits, maintains and operates its own assets, costs associated with each stage of the lifecycle are collected internally and can be accessed to build costing models.

Risk

The SLEP work has undertaken various risk assessment activities to establish the likely levels of risk associated with design options. Some of this work has required detailed engineering research and investigation in order to gain the necessary understanding to be able to assess the risks confidently.

Performance
Whatever the final design outcome, the asset must deliver on performance requirements to be an effective search and rescue asset. The key performance measures for an ALB are:

- Speed.
- Endurance.
- Sea keeping.
- Stability, inc' self-righting, survivor carrying capacity.
- Towing capability.

4.4 (b) Decision

The assessment of the new build and life extension options against the defined criteria is summarised in table 2 below:

Options / Criteria	**Life Extend**	**New build**
Cost		
Design	£2M	£3M
Build	£1.5M	£3M
Operational & Support	Similar costs	Similar costs
Risk		
Whole Body Vibration	ALARP (New crew seating)	ALARP (Same new crew seating)
Structural capability	ALARP (Validated existing structure)	Low (New design)
Electrical safety	Low (Same solution)	Low (Same solution)
Electronic obsolescence	Low (New solution)	Low (Same solution)
Daughter Boat launch & recovery	ALARP (New crane solution)	Low (potential transom door solution)
Mechanical systems	ALARP (New solution)	ALARP (similar new solution)
Propulsion	ALARP (Marine diesel solution)	ALARP (Marine diesel solution)
Performance		
Speed	25 kts	25 kts
Endurance	10 hrs	10 hrs
Sea Keeping	Meets requirements	Potentially better
Stability (inc surv' capacity)	Meets requirements	Potentially better
Towing capability	8 t bollard pull	Likely same
Strategic Outcomes		
We keep our people safe.	Safety risks ALARP	Safety risks ALARP
Focus on what will save most lives	Meets requirements	Meets requirements
Influence Safety	Same RNLI connectivity solution	Same RNLI connectivity solution
People love being part of the RNLI.	Good	Possible improved wheelhouse layout
Spend donor's money wisely.	High value. Using current assets	Lower value - high cost of new tonnage
Our future is sustainable and secure.	Re-use of hull Diesel propulsion.	New hull - same materials as current lifeboat construction. Similar diesel propulsion

Table 2. Decision summary

Based on the assessment above, the RNLI has decided that the option to life-extend the Severn class lifeboat is the better overall solution.

5. WHAT IS THE SCOPE OF THE LIFE EXTENSION AND WHAT WORK IS BEING UNDERTAKEN?

5.1 REQUIREMENTS GATHERING

To further define the requirements for the Severn Life Extension the opinions of key stakeholders were gathered, recorded and analysed. The key stakeholders being existing Severn crews, RNLI technical surveyors and engineers.

A conference was arranged in Cardiff in 2017 inviting crew representatives from all Severn stations to identify what they saw as potential improvements to the boat's design. This generated a long list which were then ranked by attendees into priority categories for inclusion in the scope of the project:

1. Safety enhancement.
2. Compliance with legislation.
3. Equipment obsolescence.
4. Significant:
    a. improvement in operational capability.
    b. reduction in through life cost.
    c. reduction in maintenance – compatibility with 12-year refit schedule (from current schedule of 7 years).

Each alteration carried out to the boat fits into one of the seven categories above.

On completion of the analysis exercise, design engineers compiled technical specification documents detailing the planned scope of the project for agreement of all key stakeholders prior to work commencing.

5.2 SCOPE OF DESIGN WORK

Whilst not a new design, the SLEP design work has attempted to address all necessary aspects of the boat to ensure that it will be capable of serving as a lifesaving platform fit for the next 25 years. The scope includes:

- Human factors, including ergonomics, noise and shock mitigation.
- Structural assessment, repairs and upgrades, including removal of aluminium fan casings.
- Verification of stability capacity.
- Significant changes to mechanical systems
- Development of RNLI Systems and Information Management System (SIMS) and communications systems.
- Better lighting and provision of HVAC system.
- Improved navigational and search lighting.
- Lighter, more responsive and less maintenance intensive daughter boat, with more reliable and safer launching crane.
- New electrical system throughout.
- Hydraulic emergency steering system.
- Changes to wheelhouse and survivor space layouts to accommodate the new equipment.

Some of these are covered in more detail below.

5.3 HUMAN FACTORS

5.3 (a) Noise and Whole Body Vibration

The existing vessel is known at times to expose the crew to high levels of noise and WBV. At time of construction the Severn class adopted sprung seating from the off-road vehicle sector to mitigate the impacts seen by the crew. This was a first for the RNLI and a direct result of moving to higher performance vessels. Additional seat development and research [12] since has kept the RNLI at the forefront of this field and the life extension provides the opportunity to insert not only shock mitigating seating (see below), but also consider cushions in survivor spaces, seat performance monitoring and helm feedback to better balance the risks of vibration exposure and service delivery.

Noise in the wheelhouse has been quantified with medium exposure levels. An extensive measurement programme has determined that improvements are required to reduce the noise exposure and improve communication and situational awareness when operating the vessel. Fitting of sound insulation is not practical due to the layout of the boat and weight constraints however a new intercom fit will allow crew to communicate effectively while simultaneously reducing noise exposure even in the highest noise environments.

5.3 (b) Ergonomics and Anthropometrics

Since the Severn class' original design work in the 1990's, the discipline of human factors has significantly gained traction. It is now viewed as vital to producing successful outcomes within all design work and a focus on the users was ensured throughout the life extension project.

In general, the focus of human factors work at the RNLI is on protecting the crews, and those they rescue, as well as ensuring those who manufacture and maintain the boats can do so safely. This focus is particularly applicable to the design of the wheelhouse, which will host the crew for the majority of each shout.

As well as applying industry-standard techniques to the human factors wheelhouse work, the RNLI has developed in-house digital human modelling (DHM) capabilities. This human model bridges the gap between non-poseable digital human models and more advanced commercial solutions.

A combination of DHM and other technical methods were used to ensure that 95% of the population would be included in design work. It is important to note the remaining 5% would not be excluded by designs, just that the environment may not be optimised for them. This is an increase on the industry standard of designing for the 5% to 95% percentile user (90% of the population) and was justified due to the high-hazard nature of the environment.

The decision to focus and further invest in human factors engineering ensures the ongoing safety and improvement of RNLI vessels, and crewing can increasingly become more inclusive.

5.3 (c) Seat Selection

With the acknowledgement that being harnessed into seats within the wheelhouse is the safest location for crew during a shout, it was important the seats selected protected the crew. This was the primary goal of the seat selection process and was weighted accordingly, however overall seat requirements encompass not only shock mitigation but also:

- Crew comfort – over long shouts of potentially more than 12 hours the seat must be comfortable both dynamically and statically.
- Suit people – the RNLI aims to accommodate our diverse range of volunteers. This includes wearing a range of PPE.

- Allow crew to deliver safety critical roles from their seat.
- Meet the RNLI's requirements for cost, maintenance and security of supply.

To achieve this several seating options were considered including those already in service with the RNLI on their all-weather lifeboat fleet and commercially available marine seats.

When testing seats the input shock magnitude and duration (both influenced by the base vessel [13] [14]) and the occupant mass are all considered as all three effect the dynamic response of the seat. To incorporate all these factors seats were tested extensively at sea on a Severn class ALB in a range of sea conditions. A range of test subjects were used covering typical crew weights to ensure that the seat chosen was suitable for the majority of crew. Shock mitigation performance was assessed using a novel method developed for the trials [15]. This approach allowed the seat performance to be assessed across a range of conditions with a reduced trials envelope. The trials demonstrated one seat that clearly outperformed the others across all weight classes, but in particular with heavy crew in the worst conditions. This seat was selected for procurement.

5.4 STRUCTURAL CONSIDERATIONS

Prior to any structural modifications, each Severn will undergo a full structural survey and inspection to identify any defects or repairs needed. Laser Shearography will be used on areas of cored structure including the hull, deck superstructure and some areas of internal structure to gain a full picture of each boat's condition and allow repairs to be carried out. This work will be carried out once all existing fit-out and systems have been removed, giving good access for inspection and repairs to areas of structure not normally accessible. This inspection work will include a particular focus on areas considered to be a structural 'hot spots' as identified through the work described in section 4.3.

Each Severn will require structural modifications to allow for the new systems and items of fit-out, including:

- Reinforcement of wheelhouse sole for new crew seats.
- Addition of new bulkhead in the wheelhouse for mounting crew seats.
- New mounting points for main and emergency steering.
- Mountings for hydraulic components and bilge pumps.
- Additional internal lifting points in engine room.
- Securing points for working at height.
- Reinforcement to wheelhouse for new search light.
- Replacing aluminium fan casing/crane mounting with composite version.

5.5 STABILITY ASSESSMENT

It is vital to ensure that the life extended Severn has similar stability characteristics as its predecessor to ensure the handling capability is maintained as well as the ongoing requirement for the boat to be inherently self-righting, as is the case for all the RNLI's modern all-weather lifeboats. Therefore, as part of the design work being undertaken, the weights and centres of each component being removed and added is being closely monitored.

5.6 MECHANICAL SYSTEMS

A fundamental part of the SLEP project was the consideration of alternative propulsion solutions to replace medium speed marine diesel for emissions reasons in line with the strategic aims.
The operational requirement to achieve 25 knots for a 10-hour duration, with a need to replenish and be mission ready in a short time, is a demanding requirement for the current state of alternative solutions.
The SLEP project reviewed the capability and implications of various alternative solutions including hybrid and full electric propulsion, however none of the options considered were able to deliver the power density necessary to meet the requirement without significant weight or space implications, which would render the design impracticable.

Accordingly, the vessel will continue to be powered by the same MTU medium speed marine diesel engines being fitted to the fleet to replace the original Caterpillar 3412 engines.

It is hoped that, in due course, an alternative less environmentally impactful liquid fuel solution will be available which can be used in these current engines. The RNLI is working closely with engine suppliers and other parties to develop a suitable roadmap to the implementation of cleaner fuels for the Severn and other assets.

Many of the boats mechanical systems are being integrated via a load sensing, fully proportional hydraulic system. This new hydraulic system will provide power for the bow thruster, Engine Room fan, HVAC compressors, bilge pumps and the daughter boat crane. Powering these items via this hydraulic system removes the current engine driven PTO assembly which reduces the associated rotating machinery risks and maintenance. It will improve the performance of the systems over the previous PTO driven setup with:

- Improved bow thruster performance in normal running and single engine operation.
- Intelligent engine room fan control via SIMS stops the fan when not required and reduces onboard noise.
- Improved bilge pump performance as these can be set at a fixed speed irrespective of the engine speed.

The bilge system is being redesigned and will incorporate one main bilge pump and one general service pump (fire main and bilge back up) with PE pipework replacing corrosion prone metallic pipework.

Steering in the wheelhouse will be via a joystick on the seat arm at helm and coxswain positions. The Upper Steering Position will still have a wheel, but with electronic input and the port side wing control station will have a joystick. The emergency steering block and tackle arrangement will be replaced with a dedicated hydraulic cylinder in the steering gear space, controlled by a wheel that attaches at the helm position in the wheelhouse.

5.7 ELECTRICAL AND ELECTRONICS

The System Information Management System (SIMS) will consist of 5 screens in the wheelhouse and 1 at the USP. Control will be via trackballs mounted on the seat arms in the wheelhouse and a trackpad at the USP.

The system is designed using standard off-the-shelf components, aiming to make it as simple to install, operate and support as possible with minimum cabling. Systems include:

- Charts and radar.
- Engine monitoring.
- CCTV – HD colour cameras throughout and thermal cameras for engine room monitoring.
- VHF radio.
- Hydraulics, bilge, seawater, switchboard and fuel system monitoring and operation.
- Mission logger (including audio data).

SIMS will also deliver a new comms system that will include:

- Wired headsets at crew seats.
- External speakers and microphones.
- Wireless ear defender headset for engine room, fully integrated into main comms system.
- Wireless headsets mounted into helmets for deck ops / close range casualty ops, fully integrated into main comms system.
- User control for headset channel selection mix - intercom, radios, VHF direction finder, sat phone and alarms.

5.8 OPERATIONAL CONSIDERATIONS

Significant improvements will reduce manual handling efforts onboard in areas such as deploying and recovering the boat's anchors, operating the man-overboard A-frame and in recovering the Severn's daughter boat (using newly installed winches).

Improved lighting is also a major operational benefit:

- All lighting inside and outside the boat including navigation lights will be upgraded to LED lighting to improve light levels and spread.
- Engine room low level lighting to provide better visibility down into the bilges.
- A high-power remote-control LED searchlight mounted on the wheelhouse roof. Controlled by joystick at the coxswain's seat and from SIMS screens at other seats.

A new thermal camera is being trialled to improve searching (FLIR M364C LR Stabilised Pan & Tilt Thermal Camera).

6. PROJECT CONSIDERATIONS

6.1 FINANCE

Initial estimates were that the cost of the Severn life extension would be in the region of £1.25M per boat. Experience gained on the prototype boat, in particular the quantity of labour hours used to build the boat, has increased this estimate to between £1.4M and £1.5M per boat split equally between material and labour costs. The forecast cost of the prototype boat is approximately £4M: £2M for design and development, that includes the cost of external design consultants and OEM design, the production of tooling and the costs attributed to the internal design team and £2M for production (direct materials and labour).

6.2 INFORMATION AND DOCUMENTATION

The current Severn class lifeboat is still supported using the original drawing pack. Although in the immediate future there will be a mixture of original drawings and new design documentation, the extension programme is an opportunity to update the documentation and bring the RNLI towards a digital future.

The development of RNLI CAD standards and an asset information strategy will improve and standardise documentation used to design and maintain the assets throughout their lifecycle. Improvements to drawing structure and document identification is one of the many steppingstones the RNLI is taking to create robust and accurate digital twins for all future assets. Having the right data, in the right place, for the right people, at the right time, will allow for more efficient and effective decision making, delivery and compliance.

6.3 PROJECT PLAN

The prototype boat is due to launch early in Q3 2021 having been in the RNLI's All-Weather Lifeboat Centre for approximately 18 months, with the first production boat planned to go on-service at Aberdeen lifeboat station in mid-2022. The number of repairs required, an immature design and the Covid-19 pandemic have all been factors in delaying the launch of the prototype boat. There are still some significant risks associated with delivering the project but it is currently on track to launch the prototype boat on schedule depending on the quantity of emergent work. Following launch the boat will undergo 2 months of system commissioning followed by extensive sea trials. Weighing and inclining trials and live self-righting trial will all be scheduled.

The plan for the life extension of the Severn fleet is to create a rolling programme where boats will come away from stations, be life-extended and then return to service but at a different Severn station.

After the initial 3-4 boats have been completed, it is expected that the refit period will drop to around a year for the remaining fleet of 40 boats and 3 boats per year will be life extended.

7. CONCLUSIONS

The technologies and materials adopted in the design and construction of the new RNLI 25 knot ALBs in the 1990s has presented the RNLI with a previously unprecedented opportunity to take advantage of a potentially lower cost solution to the activity of replacing assets at end of life. With no change to fundamental operational requirements, it has been possible to consider re-using the Severn ALB hulls to deliver an upgraded modern lifesaving asset which will not only take better care of its crew, but also help deliver parts of a wide range of organisational aims for the RNLI.

Whilst it has not proved possible to achieve large gains in all aspects, such as reducing the environmental impact of the propulsion system in the short term, the overall package of improvements and re-use will deliver good value for the RNLI.

8. ACKNOWLEDGEMENTS

The authors wish to acknowledge the contributions and support from the whole of the RNLI SLEP team in design and production. Their commitment and dedication in the midst of very challenging circumstances have made the project and this paper possible.

9. REFERENCES

1. CLARK, S.D., 1997. Long Term Behaviour of FRP Structural Foam Cored Sandwich Beams (Doctor of Philosophy). University of Southampton, Department of Ship Science.

2. CLARK, S.D., SHENOI, R.A., 1998. Fatigue Considerations for FRP sandwich structures of RNLI lifeboats. Trans.Royal Institution of Naval Architect 239–257.

3. Clark, S.D., SHENOI, R.A., Allen, H.G., 1999. Modelling the fatigue behaviour of sandwich beams under monotonic, 2-step and block-loading regimes. Composites Science and Technology 59, 471–486.

4. HUDSON, F. D., HICKS, I. A. & CRIPPS, R. M., The design and development of modern lifeboats', Proceedings of the Institution of Mechanical Engineers, Volume 207, 1993.

5. ROBERTON, D.M.V. Residual life assessment of composite structures: with application to all-weather lifeboats. University of Southampton, Engineering and the Environment, Doctoral Thesis. (2015)

6. HARMAN, A., CHAPLIN N., PHILLIPS H., & AUSTEN S., 'The Structural and Stability Assessment and Subsequent Recovery of a Damaged Lifeboat', The Damaged Ship, RINA, London, 26-27 January 2011

7. PHILLIPS, H. J., PITMAN, M., BOULTON, M., MÖLEN, H. E. VAN DER & STROMENGER, K., 'Design Verification Framework for Search and Rescue Craft', SURV 9: Surveillance, Pilot & Rescue Craft, RINA, London, April 2018.

8. PRINI F, BIRMINGHAM RW, BENSON S, DOW RS, SHEPPARD PJ, PHILLIPS HJ, et al. Enhanced Structural Design and Operation of Search and Rescue Craft, Proceedings of the 13th IMDC International Marine Design Conference, Helsinki, FI: Taylor & Francis Group; 2018, p. 439–52.

9. PRINI F. Enhanced Design and Operation of Small High-Speed Craft. PhD Thesis. Newcastle University, 2019.

10. PRINI F, BENSON S, DOW RS, PHILLIPS HJ, SHEPPARD PJ, BIRMINGHAM RW. Model Scale Prediction of Seakeeping and Global Bending Moment on a High Speed Craft. Engineering Structures, In Press.

11. PRINI F, BENSON S, BIRMINGHAM RW, DOW RS, FERGUSON LJ, SHEPPARD PJ, et al. Full-Scale Seakeeping Trials of an All-Weather Lifeboat, Proceedings of the SURV9

Conference: Surveillance, Pilot and Rescue Craft, London, UK: The Royal Institution of Naval Architects; 2018, p. 25–36.

12. CRIPPS R., CAIN, C., PHILLIPS, H. et al., Development of a New Crew Seat for All Weather Lifeboats, Proceedings of RINA SURV6, 2004, London, UK

13. COE T.E., DYNE S., SMITH J.N.. GUNSTON T., TAYLOR P., REES T., et al., Development of an international standard for comparing shock mitigating boat seat performance, Proceedings of the RINA Conference on Innovations in Small Craft Technology, April 2016, London UK

14. RILEY M.R., HAUPT KD., MURPHY, H., An Investigation of Wave Impact Duration in High-Speed Planing Craft, Naval Surface Warfare Center Carderock Division Report NSWCCD-80-TR-2014/26, April 2014

15. GUNSTON T. AND CLARKE A., Development, test and selection of a shock mitigation seat for an all-weather lifeboat, Proceedings of the RINA Conference on Human Factors, Feb, 2020, London UK

10. AUTHORS BIOGRAPHY

Neil Chaplin holds the current position of Head of Asset Management at the Royal National Lifeboat Institution. He is a member of the senior leadership team in the Engineering and Supply department of the RNLI. Responsible for the development and implementation of asset strategies as well as holding the role of RNLI chief engineer, responsible for technical standards of RNLI assets. His previous experience includes project manager for the RNLI Tamar class ALB design and build, Principal Naval Architect and a Staff Officer in RNLI operations.

Nick Fenwick holds the current position of Project Manager at the Royal National Lifeboat Institution. He is the Project Manager for the Severn Life Extension Project. His previous experience includes managing engineering projects for companies in the Marine, Defence and Oil and Gas sectors and time as a Senior Engineer in the RNLI.

Christopher Smith holds the current position of Naval Architect at the Royal National Lifeboat Institution. He is currently the Lead Naval Architect for the Severn Life Extension Project. His previous experience includes providing composites engineering support for production of the Shannon, assessment and repair of composite structures and co-ordinating sea trials for the assessment of noise and exposure to Whole Body Vibration.

Imogen Sands holds the current position of Human Factors Specialist at the Royal National Lifeboat Institution. She is a member of the Naval Architecture team in the Engineering and Supply department of the RNLI. Her role involves applying human factors techniques to lifeboat design work with the aim of protecting the crews, and those they rescue as well as ensuring those who manufacture and maintain the boats can do so safely.

Tom Coe holds the current position of Whole Body Vibration, Noise and Ergonomics Programme Lead at the Royal National Lifeboat Institution. He provides subject matter expertise in the continual reduction of risk to crew from WBV and Noise exposure. He previously held the position of Technical Lead for the UK MoD's Small Craft Concept Design Group and Maritime WBV Project as well as consulting for seat and boat builders and other agencies. He is chair of the BSI committee on Human Response to WBV

Federico Prini holds the current position of Naval Architect at QinetiQ. He previously worked at the Royal National Lifeboat Institution where, as part of the Engineering Office, was responsible for several structural design and analysis tasks for the Severn Life Extension Programme. Before joining the RNLI, he was Research Assistant at Newcastle University where he worked on a research (mentioned in this paper), in collaboration with the RNLI and Lloyd's Register, on the structural design and operation of search and rescue craft.

Holly Phillips holds the current position of Principal Naval Architect at the Royal National Lifeboat Institution. She forms part of the RNLI's Design Authority working group and manages the naval architecture team responsible for the structural and stability aspects of the RNLI's fleet of all weather and inshore lifeboats, lifeguard equipment and other assets. Her previous experience includes that of Project Manager responsible for the design and development of the E-Class Mark 2 lifeboats for the River Thames.

JOINT RESEARCH PROJECT AUTOPLAN

N. Erol, UZMAR, Turkey

SUMMARY

The paper introduces the joint research project AutoPlan, which was started by UZMAR, OES, FRIENDSHIP SYSTEMS, Technical University Berlin (TUB) and SVA Potsdam, under the MarTera (Horizon 2020) program. The goal of this project is the improvement of the safety and efficiency of planing crafts, by developing an Intelligent Navigation Assistance (INA) System. In the scope of the project, a new prototype vessel is going to be developed and built, based on an initial design provided by UZMAR (Turkey). This design will be parametrically modelled by FRIENDSHIP SYSTEMS (Germany) in order to optimize the efficiency, providing a state-of-the-art fuel saving design. The required simulation tools will be provided by the Technical University Berlin (Germany), which will also carry out a validation of the optimization results with model tests. In addition, TUB will develop a method to reliably predict the manoeuvring and instability behaviour of the planing crafts. The method will require virtual captive model tests, carried out with a RANS code, especially enhanced for this purpose. Based on these tests, a mathematical model will be developed, which takes into account all six degrees of freedom, allowing not only the prediction of rudder manoeuvres at planing condition, but also the assessment of certain types of instabilities.The method will also be experimentally validated with numerous model tests, which requires a novel experimental technique, that will be developed and installed at TUB. These mathematical models are then used by OES (Turkey), who will develop the INA system, which will support the decision making in navigation in order to avoid dynamic instabilities, like for example transversal instability or porpoising. At the end of the project, UZMAR will build a full scale prototype with an integrated INA system to prove the capabilities in practice. The full scale measurements will be carried out by SVA Potsdam (Germany) and TUB and include the measurement of all ship motions as well as the thrust. The project and first results of modelling and simulation will be presented.

1. INTRODUCTION

The sector of fast boats is an important and growing market of the European Maritime Industry. Taking into account only military and coast guard fast boats leads to an approximate number of 6500 vessels that are currently operated and an expected order volume of about 1400 new vessels with an equivalent value of approximately 70 billion USD [1]. According to the report [2] patrol boats world market is estimated to expand about 7.4% in the next five years, will go up 280 million US$ in 2024, from 180 million US$ in 2019.

On the other hand, autonomous ships are being developed and there are predictions that the boats will have remarkable share in the market. The market size of autonomous ships is $ 6.1 billion in 2018 and is estimated to reach $ 13.8 billion by 2030 as shown in Figure 1 [3].

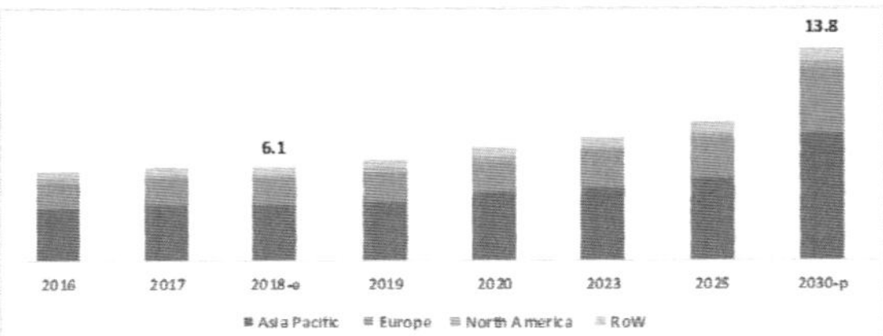

Figure 1. Market size of autonomous ships

It is estimated that 1400 new boats are ordered in addition to the approximately 6500 boats currently operated worldwide, and a trade volume of approximately 70 billion dollars will occur in the near future[4]. In addition, the autonomous ships are expected to reduce the maritime emissions and contribute to fuel savings about 20% [5].

Moreover, according to a timeline projection [6], technological developments are accelerating and it is predicted that as of 2030, autonomous ships will be tested in full scale in commercial traffic areas. Therefore, the project will be the pioneer and contribute the Turkish autonomous shipping industry.

The design of a new ship is mainly based on the classical design spiral including the hydrodynamic and structural design, the stability of the hull, the machinery and propulsion system, the safety and production limitations [7]. However, this method can be very time-consuming, therefore, more sophisticated approaches as process integration and design optimisation have been started to be applied. These approaches allow us to follow a much more efficient workflow and can even lead to better designs. Considering the entire service life of the ship from concept to disposal, the integral design process is established in order to ensure the high reliability, good performance and easy maintenance of a vessel. These three elements are achieved by a well-designed structure and good hydrodynamic performance along with the appropriate machinery system. Therefore, the rationalization and optimization of hull form is very important [8].
Apart from displacement type hulls, the high-speed planing hulls face with the longitudinal and transverse instabilities even in calm water. Longitudinal instability is based on the periodic or coupled heave and pitch oscillations called "porpoising" resulting in the bow drop and the increase of the total resistance at high speeds [9]. On the other hand, transverse instabilities end up in the large heel, the planar asymmetry and the loss of course keeping, called as "chine-walking" [10]. Along with the

increase of total resistance, these instabilities cause safety problems. According to the European Maritime Safety Agency (EMSA) [11], 20616 marine casualties happened and they resulted in 683 fatalities between the years 2011 – 2017. According to the same report, motorboats have about 5% share of all casualties and navigational incidents represent about 53.1% of all casualties. In case the causes of the accidents are analyzed in detail, it is observed that the human factor plays a significant role. Antão and Soares [12] compared the reasons for 81 accidents of ocean-going vessels and high-speed crafts and they reported that more than 50% of the accidents are based on human error. Therefore, many studies have been published about the calculation and prediction of hydrodynamic parameters and, the external and autonomous systems have been developed such as trim tabs, interceptors and electronic assistants to minimize the risks and to control the instabilities for safe and efficient navigation.

Early studies about the prediction of planning hull performance have been published in 1950s. Savitsky and Ross [13] and, Savitsky and Neidinger [14] investigated the potential whisker spray drag component of planning surfaces experimentally without the analytical derivation. Savitsky [15] investigated the viscous and pressure drag components in the bottom area and behind the stagnation line. Clement and Blount [16] conducted experimental tests to reveal the resistance of a systematic series of planing hull forms. Seo et al. [17] investigated the resistance and sea keeping performance of a mono hull experimentally. Through the decades, experimental results with the analytical approaches have been published, however, the towing tank tests are still expensive and the developed regression methods are applicable only for simple hull forms [18]. Instead of using experimental or semi-empirical methods, numerical tools have been used to predict the hydrodynamic parameters and to conduct the performance optimization of the planning hull forms.

Nowadays simple potential flow codes are increasingly replaced by more accurate viscous flow codes [19]. A common problem when performing optimizations with viscous flow solvers is the dependency of the result on the numerical grid, which alone usually leads to uncertainties in the range of 2-5%. Although this level of accuracy is sufficient for the absolute value of the target function, random errors of that magnitude would be highly detrimental for the optimisation process, since the optimisation potential itself often shows the same order of magnitude. It is therefore of key importance to ensure that grid dependencies are as small as possible and that the ranking of design variants is not impaired by the quality of the simulation. The meshing tool GridPro allows, with its approach to separate geometry and topology information, to generate rather consistent meshes with only small differences between each design. However, for larger geometry variations or variations in complex regions as for example the propeller shaft and brackets for twin-screw ships, there are still problems with this approach. Another drawback of using this approach is the time and experience required to generate the topology of the grid. A prerequisite for the automatic optimisation of geometries is the robust generation of different geometry variants. This can be achieved by a parametric hull form model, which – as opposed to the classical mathematical definition of free-form surfaces – uses design variables to control the shape of the hull. Such parametric models can be based on complex surfaces like the meta surface [20]. The software CAESES® developed by FRIENDSHIP SYSTEMS couples such parametric models with optimisation tools as for example the Sobol algorithm for a quick design space exploration or the optimisation routines from the Dakota package [19], [21].

Hydrodynamic optimizations require a considerable number of simulations to be undertaken, typically at the order of the square of the number of free variables, i.e., the degrees-of-freedom of the parametric model. Different approaches have been introduced recently to reduce the dimensionality of design spaces, most interestingly a design space reduction on the basis of Karhunen-Loève expansion (KLE) [22] – [23]. In addition to optimizing ship hull forms primarily with respect to resistance and propulsion, intensive R&D efforts are presently undertaken, primarily at the European level, to integrate as many key performance aspects as possible in advanced synthesis models [24], [25]. Due to the complexity of each simulation involved the first-principle codes are replaced in the early design phases by means of surrogate models [24], [26] for faster speed and easier handling by naval architects with less experience in simulation. In order to build high-quality surrogate models, the design space needs to be thoroughly explored in advance by means of Designs-of-Experiment (DoE). For a practical surrogate model, it is of utmost importance to not only provide a large number of geometric variants but also to investigate many operating points, stemming from combinations of trim, draft and speed. Again, a fully automatic and high-quality generation of variants and numerical grids, as well as an accurate flow analysis, are needed.

Doustdar and Kazemi [18] analyzed the effects of fixed and dynamic mesh methods for CFD simulation of a planing craft. They stated that even though the dynamic mesh method is more accurate its computational cost is higher than that of fixed method. Caponnetto [27] analyzed the hydrodynamic parameters of planing crafts and conducted steady simulations using the RANS method. Su et al. [28] investigated the hydrodynamic performance of a planing vessel using the VOF method and they found that the hydrodynamic performance of vessels at high speeds can be predicted accurately. Avci and Barlas [29] analyzed the total resistance behavior of a high speed hull using both the experimental and numerical methods. They revealed that the numerical ventilation problem (NVP) must be resolved by applying the phase replacement at high speeds. Kohansal et al. [30] calculated the hydrodynamic characters of planing hulls using a mathematical model including boundary element method and boundary layer theory. Brizzolara and Serra [31] investigated the flow characteristics around the high speed planing hulls using finite volume method. Ikeda et al. [32] investigated the transverse and longitudinal instabilities of high speed crafts in calm water. They revealed that porpoising of the craft is self-exciting based on the coefficients between pitch and heave motions. Kim et al. [33] investigated the high speed planing hulls to improve their seakeeping performances for safe navigation.

On the external systems, there are many studies on interceptors and tabs at the stern for decreasing the instabilities. Tsai et al. [34] investigated the effect of systems including stern flap, interceptor and integrated interceptor with stern flap on the resistance reduction. Pelaez et al. [35] compared the stern flaps and interceptors in hydrodynamic point of view. Karimi et al. [36] analyzed the effects of interceptors on high speed planing crafts to investigate the hydrodynamic quality and two models of planing mono-hull and catamaran were tested. Mansoori and Fernandes [10] investigated the effects of hydrodynamic interceptors by using numerical methods and Savitsky porpoising theory. They claimed that an interceptor can completely overcome the instability problems.

Even though the numerical methods are more reliable and faster compared to their predecessor applications, due to breaking waves, spray generation at hard chines and missing viscous damping of non-linear roll motion, potential theory-based codes are not precise enough for the complex problems. On the other hand, the use of RANS codes for this purpose is complicated and only a few applications can be found. As given, most of the published literature is still focusing on resistance simulations [37] and [38] apply their in-house RANS codes for calculating the running attitude and the resistance of fast planing boats, achieving a satisfactory agreement with experimental results [39]. In numerical simulations, RANS and LES are employed and compared to study the fluid field with different moving grid techniques, i.e. overset and morphing grids. At present, just a few applications for maneuvering and especially for detecting instabilities of high-speed vessels using RANS simulations are published. Villa et al. [40] predict the maneuverability of a planing craft based on simulations using the commercial code STAR-CCM+ in 4+2 DOF and obtain promising results for turning circle tests. Cura et al. have studies [41] – [42] and developed in-house codes Neptuno and Xnavis on instability and maneuverability analyses.

2. THE MODEL

The model is chosen as a planing hull considering the target of the boat. A multipurpose open deck would lead to use the hull as a pilot boat, search and rescue boat, ambulance boat including towing facilities. First of all a benchmark study for the hulls between 12 m and 24 m was prepared including the length, beam, draught, engine power, propulsion, weight information. Then the necessary ratios are calculated. A hull with a length of approximately 12 m was found to be eligible for the project's aim. Regarding the speed, it is found to be feasible to use conventional propulsion system with two propellers however the placement of the engine in a suitable angle resulted in the introduction of shaft tunnels. The speed ranges taken into account and minimum 25 knot criteria having the Froude number 1.18 is chosen. For a mothership, considering the current trend of flat stern and flat bow, the first hull lines are proposed and following the assumed weight, the chine is placed at a suitable draught. The model was prepared in Rhinoceros® Program enabling to interactively form three-dimensionally curved surfaces easily.

Type	L	B	T	DISP	Speed (	Fn	Propuls	Total e	no of e	Crew	Hull	class	L/B	B/T	L/DISP
PILOT BOAT	11,50	3,81	1,30	12500,00	24,00	1,16		633,00	2	6		BV	3,0	2,9	0,5
pilot	11,50	3,85	1,30	12500,00	29,00	1,40		745,00	2	6		BV	3,0	3,0	0,5
patrol	12,00	3,50	0,67	8750,00	35,00	1,66	waterjet	848,00	2	15	aluminum		3,4	5,2	0,6
pilot	12,00	4,20	1,20	14000,00	27,00	1,28		522,00	2				2,9	3,5	0,5
pilot	12,00	4,20	1,20	14000,00	32,00	1,52	conv	1066,00	2	8			2,9	3,5	0,5
patrol	12,40	3,60	0,60	6800,00	38,00	1,77	waterjet	551,00	2				3,4	6,0	0,7
pilot	12,80	4,30	1,30	16500,00	23,00	1,06		745,00	2	6		BV	3,0	3,3	0,5
pilot	12,95	4,00	1,20	15900,00	26,00	1,19	waterjet	633,00	2	6	grp	LRS SSC	3,2	3,3	0,5
patrol	13,40	3,80	0,70	9000,00	45,00	2,02	waterjet	894,00	2	8			3,5	5,4	0,6
patrol	13,60	4,20	0,90	17000,00	43,00	1,91	waterjet	1230,00	2	6	aluminum	Lloyds SSC	3,2	4,7	0,5
patrol	13,75	4,00	0,75	12300,00	40,00	1,77	conv	986,00	2	4	aluminum		3,4	5,3	0,6
pilot	14,00	4,60	1,40	16000,00	23,00	1,01		440,00		10	aluminum		3,0	3,3	0,6
pilot	14,20	4,10	1,32	17500,00	24,00	1,05	conv	745,00	2	8			3,5	3,1	0,5
pilot	14,30	4,10	1,35	17500,00	24,23	1,08		746,00	2	8		BV	3,5	3,0	0,6
pilot	14.95	4.50	1.34	22900.00	27.00	1.15		735.00	2	8	aluminum		3.3	3.4	0.5
pilot	15,40	5,10	1,30	17000,00	27,00	1,13		671,00	2	7			3,0	4,3	0,6
patrol	16,10	5,00	1,50	24000,00	25,00	1,02	conv	970,00	2	14		LR	3,2	3,3	0,6
pilot	16,15	5,13	1,42	23133,00	28,00	1,14	conv	1043,00	2	7	aluminum		3,1	3,6	0,6
pilot	16,20	5,10	1,20	18500,00	27,00	1,10		745,00	2	7			3,2	4,3	0,6
patrol	16,50	5,20	0,80	16000,00	45,00	1,82	waterjet	1193,00	2				3,2	6,5	0,7
pilot	16,50	5,30	1,30	24000,00	24,00	0,97	conv	894,00	2	9	grp	LRS SSC	3,1	4,1	0,6
patrol	12,60	3,90	0,82	11200,00	45,00	2,08	waterjet	838,00			frp		3,2	4,8	0,6
patrol	12,60	3,80	0,80	15000,00	35,00	1,62	waterjet	678,00	2	2	aluminum	Special se	3,3	4,8	0,5
patrol	15,70	5,40	0,80	20000,00	35,50	1,47	waterjet	1000,00	2	2	aluminum	BV	2,9	6,8	0,6

Figure 2. Benchmark study

UZMAR	UZMAR GEMİ İNŞA SAN. VE TİC. A.Ş.	Document No:	DZY-ARG-BENC-001
		First Issue Date :	8.09.2020
		Revision No:	0
	Benchmark for pilot and patrol boats	Project No:	MARTERA
		Yard No:	
		Prepared By :	Nalan EROL
Boat type	pilot		
Hull length (m)	12		
Max speed (knts)	25		
Fn	1,19		
B (m)	3,96		
T (m)	1,11		
Disp (kg)	11210,74		
Engine power (kW)	780,99		

Figure 3. Benchmark study-Excel form

Figure 4. Autoplan design image

3. PARAMETRIC MODELLING

When developing a hull form for high performance with respect to several key performance indicators it is important to be able to vary and adjust the shape with high efficacy. For a small to medium-size planing hull calm water resistance and propulsion efficiency as well as seakeeping and maneuvering are to be considered concurrently, trying to avoid instabilities (as discussed above). Consequently, a sophisticated parametric model is needed.

In principle, two types of parametric models are distinguished: (i) Fully-parametric models in which the entire shape is defined via parameters from scratch and (ii) partially-parametric models in which only the modifications induced to a baseline are defined via parameters. Both types can be nicely combined to form hybrid models that are even more powerful. Thorough discussions can be found in Harries et al. [43] and in Harries [44] .

A first parametric model was developed for the project within CAESES on the basis of a fully-parametric model originally built by FRIENDSHIP SYSTEMS for the design and optimization of a small hard-chine motorboat, see Albert et al. [45] The fully-parametric model was modified in order to capture the essence of the hull form proposed by Uzmar, see Figure 5. An interactive parametric fit was undertaken so as to start with a parametric hull that is close to the mothership without necessarily being an exact replica, see Figure 6. Since the hull form shall be optimized at a later stage of the project, and consequently will undergo smaller and larger modifications, a perfect fit was not required. Also, some of the elegant features of the freeboard were omitted for the time being. This is because the parametric model shall be used primarily for simulations and too many details would necessitate subsequent defeaturing.

Figure 5. Mothership as interactively modelled in Rhinoceros®

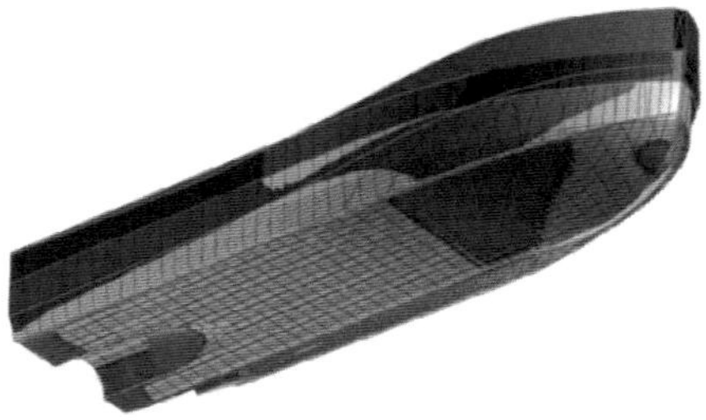

Figure 6. Comparison of mothership and parametrically fitted hull form in CAESES®

Figure 7 shows the fully-parametric hull form without tunnel. Figure 8 displays the hull form with a tunnel imprinted. The tunnel for the two shaft lines and propellers are needed to accommodate reasonably large propellers at a small enough shaft inclination while avoiding a draft that would be too deep and too low below the keel line.

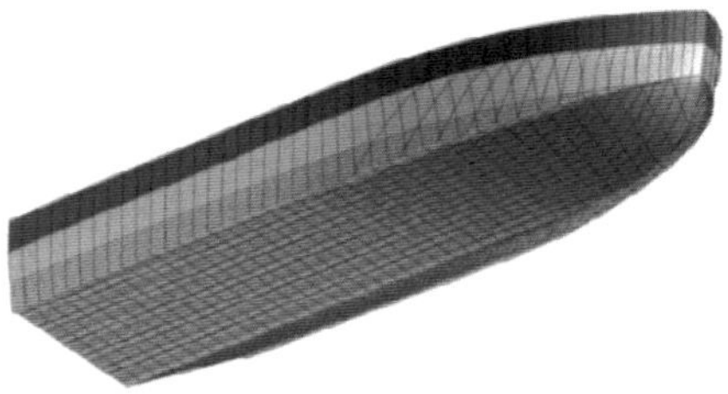

Figure 7. Fully-parametric hull without tunnel

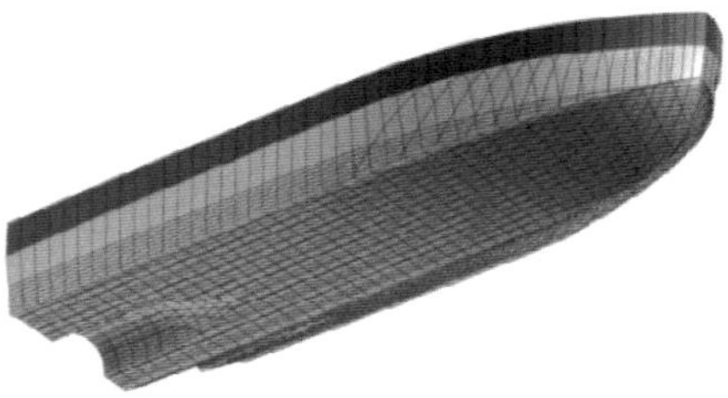

Figure 8. Hybrid parametric hull form with tunnel

Figure 10 illustrates how the fully-parametric hull form without any tunnel is modified to feature a tunnel. Here a so-called Cartesian shift is utilized which means that a surface describing the tunnel contour (shown in grey on the right side of Figure 10) is superimposed with the original hull form. For details about Cartesian shifts see Harries et al. [43].

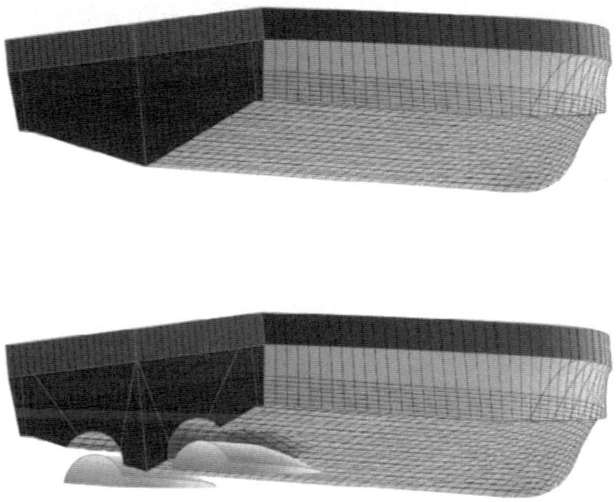

Figure 10. Introduction of tunnel via a Cartesian shift (rotated by deadrise angle at transom)

The surface defining the Cartesian shift itself is again given as a fully-parametric surface with parameters that control the radius of the tunnel, the transverse and vertical position of the propeller axis, the length of the tunnel, certain angles at the tongue (i.e., the transition area between the plain underwater part of the hull without any tunnel and the tunnel itself) etc. Several modifications are shown in Figure 11 and 12. (The variations are rather pronounced in order to better see differences.)

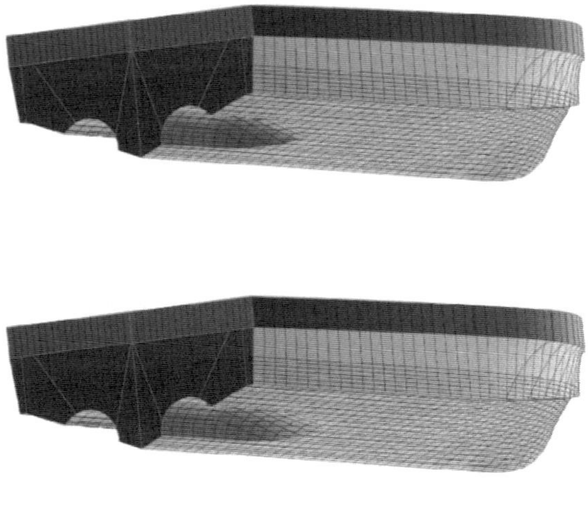

Figure 11.Parametric hull as fitted to mothership (top) and with elongated tunnel (bottom)

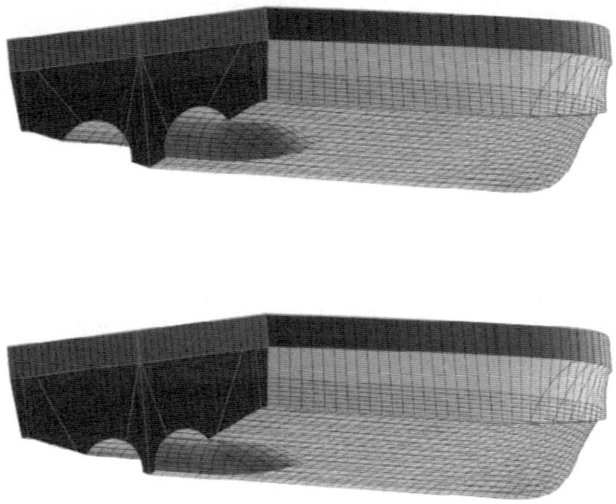

Figure 12. Parametric hull with larger tunnel radius (top) and with a pronounced inward shift of the tunnel, using the original tunnel radius (bottom)

Since the hull form without tunnel itself is controlled by parameters, too – for instance main parameters such as beam, draft and length of the ship as well as detailing parameters describing important features such as the deadrise angle at the transom, the breadth of the spray rails etc., see Albert et al. [45] – the hybrid model allows flexible variations at a wide scale. Figure 13 and 14 illustrate modification stemming from a change of beam and a change of the deadrise angle at the transom. Here the tunnel geometry was kept but, naturally, any parametric alteration as shown above could be superimposed.

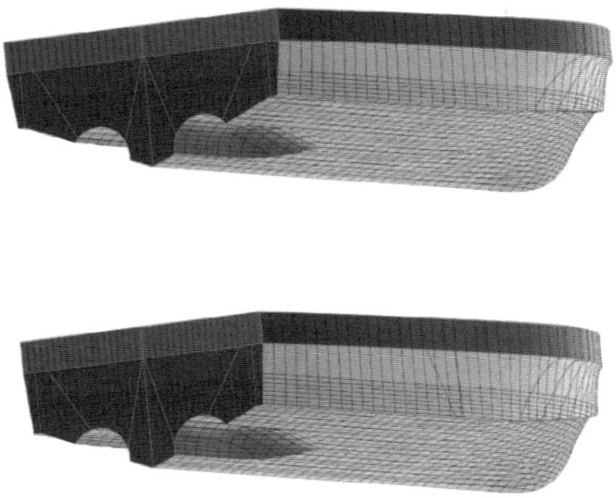

Figure 13. Parametric hull form with enlarged (top) and reduced beam (bottom)

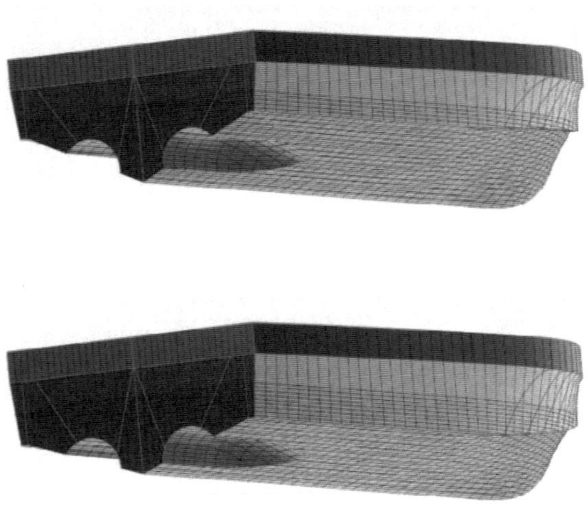

Figure 14. Parametric hull form higher (top) and lower deadrise angleat transom (bottom)

4. DEVELOPMENT OF MANEUVERING PREDICTION TECHNIQUE

In order to achieve a high level of safety and efficiency when building a planing hull, it is crucial to determine the manoeuvring characteristics as well as the required power already during the design phase. Moreover, a prediction of manoeuvring and especially of unstable running attitudes will enable the Intelligent Navigation Assistance System mentioned below to foresee critical situations. The key research aspect of the present project therefore is the development of a reliable manoeuvring prediction technique, improving the state of art regarding planing crafts.

The intended method will take into account all six degrees of freedom, since, in contrast to displacement hulls, planing crafts experience large changes in their running attitude during manoeuvres, which in turn significantly influence the manoeuvring behaviour. The hydrodynamic forces acting on the hull will be calculated by means of a system based mathematical model of Abkowitz type. The hydrodynamic coefficients are determined throughout virtual captive model tests, which will be conducted with the RANS code Neptuno, see Cura Hochbaum et al [46], that will be enhanced especially for this purpose in order to accurately allow the prediction of the running attitude of the vessel. Solving the motion equations in six degrees of freedom with help of the mathematical model yields then the trajectory and all motion parameters during any desired manoeuvre in fractions of a second. If considering the hydrodynamic coefficients for the dependency on the rudder angle and all motion parameters, including velocities and accelerations as well as all possible combinations of these, the quantity of required virtual captive model tests would reach exorbitant dimensions that are not practicable. The number of coefficients involved in the mathematical model therefore has to be delimited to a convenient extent. Different approaches will be explored during the research project. A promising approach is a 4+2 degrees of freedom method, e.g., according to Yasukawa [47], that will be further developed. In this approach, the hydrodynamic coefficients are deployed for four degrees of freedom, but in turn are a function of the remaining two. This procedure reduces the number of coupling terms and thus the number of required virtual captive model tests.

The motion equations and hydrodynamic coefficients will be employed for a linear stability analysis regarding typical instabilities of planing crafts, like dynamic loss of transverse stability, heaveand pitch instability (porpoising) and yaw instability. Stability indices will be determined, allowing a very quick evaluation of the stability of the current running attitude of the vessel.

All developed theoretical numerical methods will be validated via model tests that will be conducted at TUB's experimental facilities as well as by means of full scale measurements performed with the prototype. The experiments, with the model being free to rise and trim, will contain resistance and propulsion tests, rudder force measurements at propulsion condition as well as several oblique static captive model tests. In order to determine the roll moment of the vessel, some of the tests will be conducted at different static heeling angles. To implement that, an innovative testing device will be developed an installed at TUB.

Bringing together the developments of numerical, mathematical and experimental techniques, a considerable step forward to the prediction of manoeuvring and instabilities of planing craft is expected within this research project.

5. INA SYSTEM

To increase the safety during operation of the craft, an intelligent navigation assistance (INA) system will be designed and developed by Offshore Engineering Solutions Inc. (OES) in the scope of this project. The INA system will receive information from sensors and systems on board such as gyrocompass, GPS, Radar (X-band), fusion sensors, satellite compass and INS (inertial navigational system). Using this information and mathematical models developed by TUB, navigational algorithms will be created by OES. According to the results of this algorithm, the engines, interceptors and rudder of the craft will be managed. At the same time, the craft will be able to do automatic route planning, guidance and navigation with the artificial intelligence algorithm to be integrated into the system. With these algorithms, INA system will foresee and prevent critical situations, eg. By limiting the rudder angle in certain situations. This INA system will also have a power management system that will instantly supply the required energy to the system by automatically managing the engines, batteries and uninterruptable power supplies. For some situations an

automatic stabilisation using active control mechanisms may be employed. Thus, the system will provide the craft with autonomous navigation capability. The INA system thus allows for example a reduction of the bridge crew or an increase defficiency of the crew by reducing the crew fatigue.
The efficient data collection in a single system reduces the number of interfaces the crew has to manage and thus further increases the overall safety of the vessel. This may allow long-termmissions, e.g. for search and rescue operations in the Mediterranean Sea. The integration of the INA system into the prototype allows a thorough validation of its capabilities and an improvement of the used algorithms.

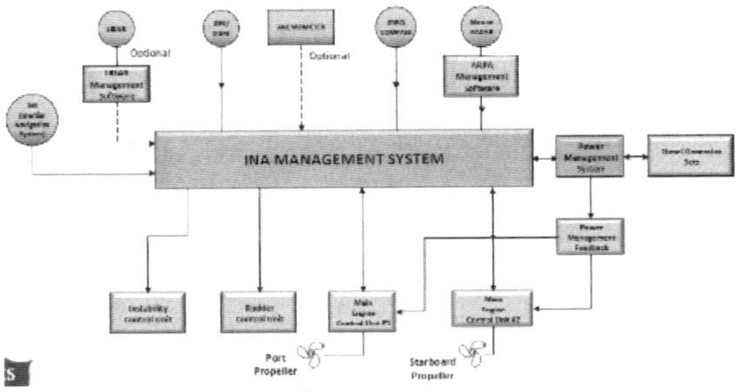

Figure 16 INA System layout

6. FIRST RESULTS

As the work has just started, only the preliminary design and preliminary paratmetric modeling are present. It can be foreseen that with the appropriate propeller and current engine configuration it is possible to reach the design speed. However in order to be able to decide on the final design, the tunnels, the chines etc are to be optimized together with considering the manuvering stability.

7. CONCLUSIONS

The joint work of the partners offers a valuable contribution to both autonomous shipping and planning craft stability as well as developing an innovative testing device. It is believed that having mathematical, parametric models and the test models will make this work valuable also for the academic literature.

8. REFERENCES

[1] Nugent B. 2019. Große Kampfschiffe - Aktuelle Vorhaben. MarineForum, 7/8.
[2] 360 Research Reports. 2019. Global patrol boats market 2019 by manufacturers, regions, type and application, forecast to 2024.
[3] Market Reports. 2018.https://www.marketsandmarkets.com/Market-Reports/autonomous-ships-market-267183224.html.Accessed on May. 2020
[4] Nugent B. 2019. Große Kampfschiffe - Aktuelle Vorhaben. MarineForum, no. 7/8-2019.
[5] Research and Markets. 2020. Autonomous Ships Market - Growth, Trends, and Forecast.
[6] DIMECC. 2017. Timeline for autonomous ships. https://news.cision.com/tieto/i/dimecc-timeline-for-autonomous-ships,c2148368. Accessed on May. 2020.
[7] BertramV. and Schneekluth H. 1998. Ship Design for Efficiency and Economy. 2. Ed., Butterworth-Heinemann.
[8] Biran A. and Pulido R.L. 2013. Ship Hydrostatics and Stability. 2. Ed., Butterworth-Heinemann.
[9] Xi H, Sun J. 2006. Feedback stabilization of high-speed planing vessels by a controllable transom flap. J Oceanic Eng. 31(2), 421–31.
[10] Mansoori M., Fernandes A.C. 2016. The interceptor hydrodynamic analysis for controlling the porpoising instability in high speed crafts. Applied Ocean Research.57, 40 – 51.
[11] Annual Overview of Marine Casualties and Incidents. 2018. European Maritime Safety Agency.
[12] Antão P., and Soares C.G. 2008. Causal factors in accidents of high-speed craft and conventional ocean-going vessels. Reliability Engineering and System Safety, 93, 1292–1304.
[13] Savitsky, D., and Ross, E. W. 1952. Turbulence Stimulation in the Boundary Layer of Planing Surfaces-Part II, Davidson Laboratory, Stevens Institute of Technology, Hoboken, NJ, Report No. 444, August.
[14] Savitsky, D., and Neidinger, J. W. 1954. Wetted Area and Center of Pressure of Planing Surfaces at Very Low Speed Coefficients, Davidson Laboratory, Stevens Institute of Technology, Hoboken, NJ, Report No. 493, July.
[15] Savitsky, D. Hydrodynamic design of planing hulls. 1964. Marine Technology, 1.
[16] Clement E. P. and Blount D. L.1963. Resistance Tests of a Systematic Series of Planing Hull Forms, SNAME Transactions.
[17] Seo J., Choi H.K., Jeong U.C., Lee D.K., Rhee S.H., Jung C.M., Yoo J. 2016. Model tests on resistance and seakeeping performance of wave-piercing high-speed vessel with spray rails. International Journal of Naval Architecture and Ocean Engineering 8, 442 – 455.
[18] Doustdar M.M., Kazemi H. 2018. Effects of fixed and dynamic mesh methods on simulation of stepped planing craft. Journal of Ocean Engineering and Science. 4, 33–48.
[19] Harries S. 2018. Upfront CAD – Parametric modeling techniques for shape optimization in Advances in Evol. and Determ. Methods for Design, Optimization and Control in Engineering and Sciences. Glasgow, Springer.
[20] Harries S. Practical Shape Optimization using CFD. www.friendship-systems.com. Accessed on Nov. 2014.
[21] Albert S. et al. 2016. Hydrodynamic Optimization of a Power Boat in the Cloud.
[22] Diez M., Campagna E. F. and Stern F. 2015. Design-space dimensionality reduction in shape optimization by Karhunen–Loève expansion. Comput. Methods Appl. Mech. Engrg., vol. 283.
[23] Bergmann E. et al. 2018. Massive Parameter Reduction for faster Fluid-dynamic Optimization of Shapes. Int. CAE Conf. and Exhibition, Vicenza, Italy.

[24] Harries S. and Abt C. 2019. CAESES – The HOLISHIP Platform for Process Integration and Design Optimization. A Holistic Approach to Ship.Design – Vol. 1: Optimisation of Ship Design and Operation for Life Cycle. Springer.
[25] Harries S. et al. 2019. Approach to Holistic Ship Design – Methods and Examples. COMPIT. Tullamore, Ireland.
[26] Will J. and Most T. 2009. Meta model of optimized Prognosis (MoP). ANSYS Conf. & 27th CADFEM Users Meeting. Leipzig, Germany.
[27] Caponnetto M. 2001. Proceedings of the Second International Euro Conference on High Performance Marine Vehicles, HIPER. 01, 128–138.
[28] Su Y., Chen Q., Shen H., Lu W. 2012. Numerical Simulation of a Planing Vessel at High Speed. J. Marine Sci. Appl. 11, 178-183.
[29] Avci A.G. and Barlas B. 2018. An experimental and numerical study of a high speed planing craft with full-scale validation. Journal of Marine Science and Technology. 26, No. 5, 617-628.
[30] Kohansal A.R., Ghassemi H., Ghasi M. 2010. Hydrodynamic characteristics of high speed planing hulls, including trim effects. Turkish J. Eng. Env. Sci. 34, 155 – 170.
[31] Brizzolara, S., Serra F. 2011. Accuracy of CFD codes in the prediction of planing surfaces hydrodynamic characteristics. 2nd International Conference on Marine Research and Transportation, Italy.
[32] Ikeda Y., Katayama T., Tajima S. 1998. A prediction method of vertical lift component of the roll and heave damping. J Kansai Soc. Naval Arch. Jpn. 229, 105–10.
[33] Kim, D. J., S. Y. Kim, Y. J. You, K. P. Rhee, S. H. Kim and Y. G. Kim (2013). Design of high-speed planing hulls for the improvement of resistance and seakeeping performance. International Journal of Naval Architecture and Ocean Engineering 5, 161-177.
[34] Tsai F.J., Hwang J.L. 2004. Study on the compound effects of interceptor with sternflap for two fast monohulls. MTTS/IEEE Techno-Ocean. 2. 1023–8.
[35] Peláez G., Martín E., Lamas A.M., Ulloa A.F., Prieto D. 2010. Preliminary study of a newstern device to improve efficiency in a fishing vessel. First International Symposium on Fishing Vessel Energy Efficiency.
[36] Karimi M.H., Seif M.S., Abbaspoor M. 2013. An experimental study of interceptor's effectiveness on hydrodynamic performance of high-speed planing crafts. PolıshMarıtıme Research. 2 (78), 21-29.
[37] Fu T.C. et al. 2012. A Detailed Assessment of Numerical Flow Analysis (NFA) to Predict the Hydrodynamics of a Deep-V Planing Hull. 29th SNH. Gothenburg, Sweden.
[38] Durante D. et al. 2015. Performances Analysis of a Semi-Displacement Hull by Numerical Simulations. MARINE.
[39] Marco A.D., et al. 2017. Experimental and Numerical Hydrodynamic Analysis of a Stepped Planing Hull. Appl. Ocean Research. 64.
[40] D. Villa et al. 2016. Application of CFD Calculations for the Improvement of Planing Crafts Manoeuvrablilty Mathematical Models. 12th Intl. Conf. on Hydrodynamics, Egmondaan Zee, Netherlands.
[41] Uharek S. and Cura Hochbaum A. 2018. Abschlussbericht des Teilvorhabens SatMotion - Satellitenbasierte Bestimmung der Schiffsbewegung. Berlin.
[42] Cura Hochbaum A. and Uharek S. 2016. Prediction of Ship Manoeuvrability in Waves Based on RANS Simulations. 31st SNH, Monterey, USA.
[43] Harries S., Abt C., BrennerM. 2018. Upfront CAD – Parametric Modeling Techniques for Shape Optimization, International Conference on Evolutionary and Deterministic Methods for Design, Optimization and Control with Applications to Industrial and Societal Problems (EUROGEN 2015)published by E. Minisci et al. (Eds.) in Advances in Evolutionary and Deterministic Methods for Design, Optimization and Control in Engineering and Sciences, Springer 978-3-319-89986-2 September 2015 (2018) | Glasgow, UK
[44] Harries S. (2020). Practical Shape Optimization Using CFD: State-Of-The-Art in Industry and Selected Trends, Conference on Computer Applications and Information Technology in the Maritime Industries (COMPIT 2020), August 2020, Pontignano, Italy
[45] Albert S., Harries S., Hildebrandt T., Reyer M.(2016). Hyrodynamic Optimization of a Power Boat in the Cloud, High-Performance Marine Vehicles (HIPER 2016), October 2016 ,Cortona, Italy
[46] Cura Hochbaum, A. and Vogt, M. (2002). Towards the simulation of seakeeping and maneuvering based on the computation of the free surface viscous ship flow. Proceedings of the 24th ONR Symposium on Naval Hydrodynamics, Fukuoka, Japan, 2002.
[47] Yasukawa, H., Hirata, N. and Nakayama, Y. (2016). High-Speed Ship Maneuverability. Journal of Ship Research, Vol. 60, No. 4, pp. 239-258.

9. AUTHORS BIOGRAPHY

Nalan Erol is a R&D Project Leader at Uzmar Shipyard, she's a shipbuilding and ocean engineer for eleven years and a PhD candidate.

WHAT NEEDED TO HYBRID? MARINE PROPULSION GREENING PRINCIPLES

M Zeid, BERG Propulsion AB, Sweden

SUMMARY

Greening and hybrid propulsion are being more and more common on the market. Apparently, looking to propulsion machinery arrangement and recognizing electric motor and batteries reflects a green propulsion system for the first while. However, being green it doesn't mean only the electric motor and batteries. The greening is mainly about energy utilization and optimization in sequence layers. This means greening concept as a start layer(s) has a direct connection to what is power configuration on board and what propulsion system available and how to utilize the available power to the specified propulsion. As a next layer(s), the optimization can be enhanced by utilizing digitalization possibilities to adopt system dynamically for the most optimum hybrid condition. The paper is going through this process where explaining optimization differences between different propulsion systems for being more and more green.

NOMENCLATURE

[Symbol]	[Definition] [(Unit)]
G	Green solution percentage (%)
Es	Sustainable Energy supplied (%)
Ens	Non-Sustainable Energy supplied (%)
Ut	Sustainable Energy utilized (%)
Utn	Non-Sustainable Energy utilized (%)

1. INTRODUCTION

Having a green solution is a major goal for all industries today. For the most, green solution is associated with electrification solutions where electricity considered as clean energy. On the other hand, the relative development on the electric motors could show up another sides of electrification advantages where the ease of use and the ease of implementation achieved. These side advantages led to even further popularity of electrification where may not fully aligned to the use of sustainable energy and (or) energy utilization. It maybe helpful to identify green solution in scaling principle rather than binary identification. This as the system can be identified in respect of sustainable energy integration value and energy utilization value where addressed to actual operation.

2. ENERGY UTILIZATION

2.1 SAILING CONDITION OPTIMIZATION

Sailing condition optimization is widely known by combinatory mode integration.

This as considering sailing speed constant, adjusting the propeller blade pitch and propeller RPM can result in significant reduction on power consumption respectively better utilization of available input power to the propulsion system.

A general layout of combinatory curve [1] is presented in Figure 1. The concept of combinatory mode is based on the utilization of changing propeller pitch in order to achieve vessel speed at relative RPM. This as reference to natural physics, higher propeller pitch at lower RPM results in same vessel speed. In order to achieve the concept, relative suitability of the propulsion equipment on board are essential where mainly a control system for pitch settings as well as propeller RPM is needed. A controllable pitch propeller is essential too. It could be some challenges for electric equipment integration, however relative solutions are available as well.

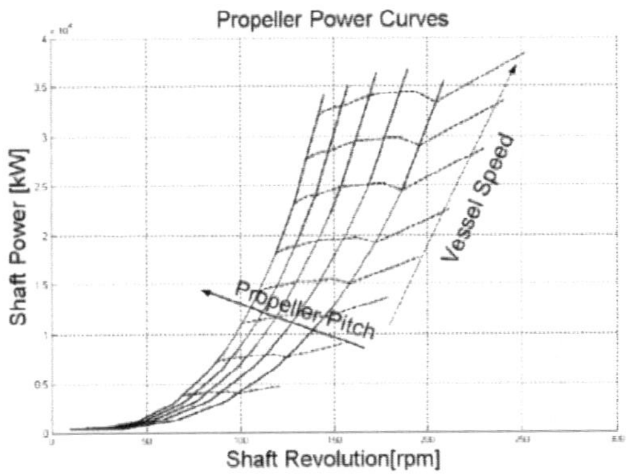

Figure 1: general layout of combinatory curve

2.2 SAILING TIME OPTIMIZATION

Sailing time optimization is one of the key power saving parameters, this as the lower vessel speed the less power consumption. Reference to propeller law at a single propeller blade pitch ration, the resulted vessel speed is not improved as much as the power consumption improvement due to the power 5 proportion ratio between the propeller power consumption and propeller revolution, see Figure 1.

For a specific sailing distance, sailing at lower speed get better utilization of power available on board if there is available time.

It used to be the applicable case for better power utilization in case of single pitch propellers.

2.3 VESSEL POWER CONFIGURATION

2.3 (a) Prime mover operational optimization

For each prime mover there is optimum operating domain. This optimum operating domain vary from prime mover to other relative to type, maker, fuel, operating speed,.....

In Figure 2, it reflects relative efficiency differences between different types of prime movers / fuel [2], regardless the accuracy of the values of curves where may not in line with today technology but the idea of potential differences between prime mover to the other is the core point.

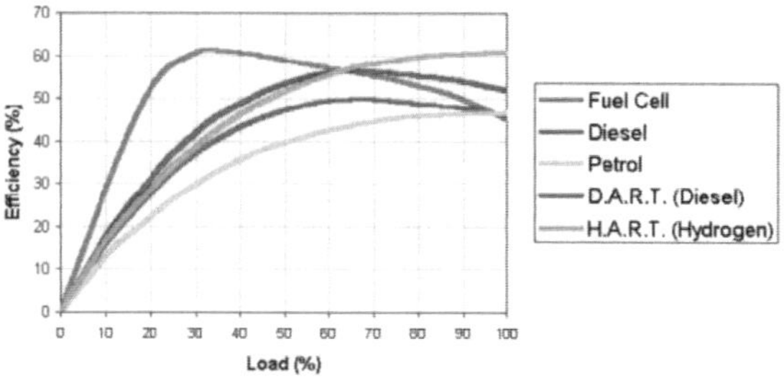

Figure 2: Relative efficiency differences between different types of prime movers

The understanding about the optimum operating domain for the prime mover is key point to either select the prime mover / fuel or to answer the question of what prime mover for propeller to use at specific operating conditions.

2.3 (b) Power configuration optimization

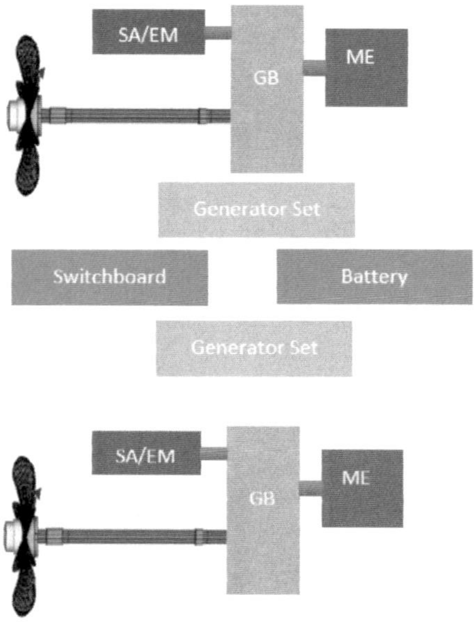

Figure 3: Power configuration possbility illustration.

The power configuration of a vessel has a major influence of power utilization of the vessel. The power configuration illustration in Figure 3 shows different possibilities for power configuration.

The major contributor of the power configuration is the vessel operating profile. Accoroding to the vessel operating profile it can be selected what is optimum number / size of propeller(s). What is more siutable prime mover(s)? is enternal compusion engine only most optimum for operation profile or hybrid the system with electric motor to cover wider operating range with high ratio of power utilization?

A lot of questions where associated with the vessel operational profile where associated with the power configuration on board.

The most important is the understanding of the importance of power configuration into improving the power utilization value of a propulsion system.

3. SUSTAINABLE ENERGY INTEGRATION

The sustainable energy integration to a propulsion system requires electrical storage (batteries) as well as sustainable electrical supply where generated by (but not limited to) solar energy, wind energy, hydro energy, tidal energy, geothermal energy or biomass energy.

As per today technology, it is still quite way to go with middle size and big size vessels. However, it may not be the same long way for smaller boats or Surveillance, Search and Rescue Crafts.

4. GREENING SCALE

Identification of a green solution is widely influenced by parameters. The major contributors as explained are energy utilization value and energy source (Author opinion).

Considering a full green solution is 100% powered by sustainable energy and utilized by 100%.

On the other side, the non-green solution is maximum when the solution is 100% powered by non-sustainable energy and not utilized at all.

Thus, the greening value in percentage can be presented as:

$$For\ Distance\ (A-B), G\,\% = \frac{(Es\,\% \times Ut\,\%) + (Ens\% \times Utn\%)}{2}$$

The greening percentage should be addressed to specific sailing distance, this as being green it is at the end choice been taken on board.

In that sense a recognition of the engineering efforts for both utilizing the available energy as well as providing solutions where based on sustainable energy supported (Author opinion).

5. CONCLUSIONS

It is always reflecting a pleasure and good hopes for future when developing engineering system that better for environment.

Sailing green is a sailing style, yes it requires specific preparation and equipment on board. However, the preparation and the equipment will not make the vessel sailing green by itself.

Identifying a specific system by a green solution is a big responsibility. However, it is always beneficial to find a quantification way to be able to know as engineers where we were, where we are and respectively where to go (Author opinion).

As per today technology, it is always possible to measure on existing systems where to identify the optimum operating conditions for better energy utilization on board.

When looking to marine propulsion systems in respect of green solutions, the following principles to consider:

- Energy utilization:
    - Propulsion efficiency.
    - Prime mover efficiency.
    - Suitable power configuration.
- Energy sources:
    - The sustainable energy sources the better where possible.

6. ACKNOWLEDGEMENTS

My deep appreciation to Royal Institution of Naval Architects for getting the opportunity to join this conference.

7. REFERENCES

1. Altosole, M., Borlenghi, M., Capasso, M.,Figari, M., 'COMPUTER-BASED DESIGN TOOL FOR A FUEL EFFICIENT -LOW EMISSIONS MARINE PROPULSION PLANT', https://www.researchgate.net/figure/Propeller-performance-prediction-curves_fig1_266576124, 12Jan2021.
2. Tomar, V. S., 'Why engine efficiency low at part load?', https://www.quora.com/Why-engine-efficiency-low-at-part-load, 2016.

9. AUTHORS BIOGRAPHY

Mohamed Zeid I'm currently holding the position of Rotordynamics Expert at BERG Propulsion AB, Sweden. I'm the global company expert for Rotordynamics related tasks, I held several positions where related to Rotordynamics activities from simulations tasks at office to troubleshooting complicated cases at shipyards since 2006. I held internal trainings within organization where resulted in several Rotordynamics specialists around the globe. Recently, I'm trying to build another team where more focused in practical measurements. As originally a Naval Architect and Marine Engineer, my previous experience includes intact / damage stability, structure analysis and strength calculations, machinery configuration and arrangements, pipelines design, etc.

FUTURE TECHNOLOGY AND PRESERVING LIFE AT SEA

J Dalziel, Dalhousie University, Halifax, Canada
Technical personnel with IMRF Future Technology Panel member organizations

SUMMARY

Drowning is one of the leading causes of accidental death worldwide. Many of the world's leading Maritime Search and Rescue organizations, members of the International Maritime Rescue Federation, work to reduce this loss of life, through both rescue response and prevention. The Future Technology Panel are representatives from a number of these organizations who share knowledge and experience in the application of various new technologies, seeking to enhance health and safety for all who venture onto the water. These include such diverse technologies as remotely piloted air and marine vehicles, automatic transcription of radio communications, advanced emergency alerting systems, search equipment and patterns, target identification and recognition, reducing the environmental footprint of rescue services and ergonomic considerations for the crew members.

NOMENCLATURE

AIS	Automatic Identification System
CCG	Canadian Coast Guard
IAMSAR	International Aeronautical and Maritime Search and Rescue Manual
IMRF	International Maritime Rescue Federation
IMO	International Maritime Organization
JRCC	(Canadian) Joint Rescue Coordination Centre
PIW	Person in the water
RHIB	Rigid hull inflatable boat, sometimes also referred to as a RIB.
RPAS	Remotely Piloted Air System – Often referred to as a DRONE, or Unmanned Aerial Vehicle (UAV).
RPMS	Remotely Piloted Marine System – Term used in this Paper for an unmanned craft, remotely controlled. Other terms, such as Remotely Controlled Rescue Device (RCRD), Unmanned Marine Vehicle (UMV), Unmanned Surface Vehicle (USV), Unmanned Underwater Vehicle (UUV), Maritime Autonomous Surface Ship (*MASS*) may be used. It is a very broad term and could potentially even involve 'marinized' Unmanned Aerial Vehicles (UAVs). All RPMS referred to in this Paper are surface craft.
SAR	Search and Rescue
SAR Tech	Search and Rescue technician, a specially trained member of an elite group of primary care paramedics that provide on-scene medical aid and evacuation, may be deployed by helicopter or parachute.

1. INTRODUCTION

Figure 1: Rescue in 1828 of 163 passengers and crew from the 'Despatch' by 17-year-old Ann Harvey, her 12-year-old brother, their father, and their Newfoundland dog 'Hairyman' (Butler) [1, 2].

1.1 THE INTERNATIONAL MARITIME RESCUE FEDERATION

"a large body of men ... in constant readiness to risk their own lives for the preservation of those whom they have never known or seen, perhaps of another nation, merely because they are fellow creatures in extreme peril." (Sir William Hillary, founder of the Royal National Lifeboat Institution (RNLI), 1823) [3]

Drowning is the 3rd leading cause of unintentional injury death worldwide, accounting for 7% of all injury-related deaths. The World Health Organization (WHO) in 2020 estimates that 320,000 people drown every year.

The International Maritime Rescue Federation (IMRF) is the international non-governmental organisation (NGO) working to develop and improve maritime search and rescue (SAR) capacity around the world, improving maritime SAR capability worldwide.

In July 1924, the Royal National Lifeboat Institution (RNLI) celebrated the centenary of its foundation (1824), by hosting the first International Lifeboat Conference in London. In announcing its plans, the RNLI stated that the conference would discuss "practically every question of technical interest, from the construction of lifeboats and lifeboat engines to the latest mechanical devices, such as tractors and line throwing guns". Eight countries in addition to the host nation were represented at the first conference, Denmark, France, Japan, The Netherlands, Norway, Spain, Sweden and the USA. It was subsequently agreed to hold these conferences once every four years, and that the RNLI would serve as the secretariat. Eventually, the International Lifeboat Federation was formed, becoming the International Maritime Rescue Federation in 2007.

The International Maritime Rescue Federation brings together the world's maritime search and rescue organisations and has both voluntary and governmental organisations as members. As well as capacity building, the IMRF also has an advocacy role, providing an international voice for its members.

The International Maritime Rescue Federation is the only maritime search and rescue non-governmental organisation with consultative status at the United Nations' International Maritime Organization (IMO).

The International Maritime Rescue Federation:

- facilitates cooperation;
- the exchange of information, research and development, advice and consultancy between maritime search and rescue services around the world;
- promotes public education and raises awareness of the importance of safety on the water.
- encourages and supports the formation and development of new maritime search and rescue services; and
- provides training in maritime search and rescue operational skills, including those required for a Mass Rescue Operation (MRO) [4].

1.2 THE IMRF FUTURE TECHNOLOGY PANEL

The Future Technology Panel of the International Maritime Rescue Federation was established by several member organizations in 2016. It currently meets monthly on Microsoft Teams, and is open to all IMRF member organisations.

The panel discussions increase participant's knowledge of emerging Search and Rescue (SAR) technologies, can improve access to research activities and expertise, and encourages cross-collaboration between organisations, reducing the cost and delay of duplication. It aims to link subject matter experts from different member organisations to share learning about SAR-related technological developments.

Past discussion topics have included collaborative ergonomic and human centred watercraft design (e.g., whole body vibration), simulation (virtual reality, training simulation), autonomous vehicles (surface and air) and modelling the maritime environment (wave prediction, etc.).

Organizations may wish to be involved in the FTP if they have a SAR topic or technology challenge they would like to examine and develop. Similarly, if they have developed new SAR systems or equipment which they want to share or discuss with subject matter experts from other IMRF member organisations [5].

1.3 SAR ORGANISATIONS

The following are the presentations from the following members of the IMRF Future Technology Panel:

1.3 (a) Organisations

- Canada – Canadian Coast Guard – CCG
- Germany – German Maritime SAR Service – DGzRS
- Norway – Norwegian Society for Sea Rescue – NSSR
- South Africa – National Sea Rescue Institute – NSRI
- Sweden – Swedish Sea Rescue Society – SSRS
- UK/Republic of Ireland – Royal National Lifeboat Institution – RNLI

1.3 (b) Companies

- Australia – Yellow Scorpion Pty. Ltd., Perth, WA.
- Republic of Ireland – SafeTrx; 8 West Consulting Ltd., Cork.

2. PRESENTATIONS

2.1 CANADA – CANADIAN COAST GUARD – CCG – EVALUATION OF REMOTELY PILOTED SYSTEMS

Sheyla Dussault, CCG National Manager SAR Policy

The Canadian Coast Guard initiated a pilot program in the Summer of 2020 to identify, evaluate and experiment operationally with remotely piloted systems capable of operating from ships or from shore. These are Remotely Piloted Marine Systems (RPMS), and Remotely Piloted Air Systems (RPAS). Several suitable devices were identified. Testing and evaluation will take place for a variety of requirements: air and marine domain data collection, direct support of maritime Search and Rescue, spill response, Vessels of Concern Surveillance, Maritime Security, Ice Reconnaissance, and Conservation & Protection Enforcement.

'Remotely piloted' makes it clear that there is a human in the loop, a person is controlling the device. But they are located remotely; perhaps nearby, on a vessel or ashore or even in a helicopter; perhaps further away in a shore

station, a mobile (truck mounted) base, or a long-range aircraft.
Data collected during testing and evaluation will be analyzed, and results documented to aid in determining future RPAS & RPMS requirements and uses to support Coast Guard programs into the future.

2.1 (a) Remotely Piloted Marine Systems (RPMS)

Remotely Piloted Marine Systems (RPMS) show potential for enhancing crew safety during the rescue phase of SAR operations. The use of flotation devices on board ships aims to assist with saving lives, especially in the case of a person in the water (PIW) as part of a SAR response. Not having proper tools has, in the past, motivated rescue crews to put themselves in unsafe situations leading, in a few unfortunate cases, to the death of rescue personnel. In other cases, the perceived inability to help victims has resulted in subsequent mental wellness issues.

The initial intent motivating the acquisition of RPMS is to gain a capability akin to a motorized Lifebuoy (Kisby ring). Lifebuoys are required by national Regulation (and by SOLAS if applicable) to be carried on board Coast Guard and other ships, and also at docking facilities, for assisting in PIW recovery operations. While aiding PIWs, Coast Guard policy directs that rescuers are not to enter the water to assist. For both the RPMS and Lifebuoy, the PIW must be conscious, and have a level of dexterity to use the device, however, for instance, the PIW may have limited hand mobility (ability to hold on to the device), such as from hypothermia. A RPMS may be fitted to offer various options for successful rescue in such cases, such as using a line or strap to wrap around one's arms and be pulled, or large handles through which one can put their arm.

The concept of operations for the SAR RPMS includes the deployment of devices from:

- Shore. A qualified operator will launch the device from a shore location, either water-level or dropped/lowered from an elevated position (bridge, lock, cliff, etc.) in support of a rescue mission or vessel crew overboard. The authority to launch the device will rest with the station or vessel.

- Vessel. A qualified operator will launch the device from a vessel in support of a JRCC rescue mission or vessel crew member overboard. The RPMS can also be used to bring a line or other light equipment (radio, PFD, helmet) to another vessel or a beach/shore where other rescuers are participating in the response. The authority to launch the device will rest with the vessel Captain.

Three suitable devices were identified and evaluated (with some early input from the NSRI of South Africa). Testing of three selected devices is has commenced in early 2021, with the test report expected in the summer.

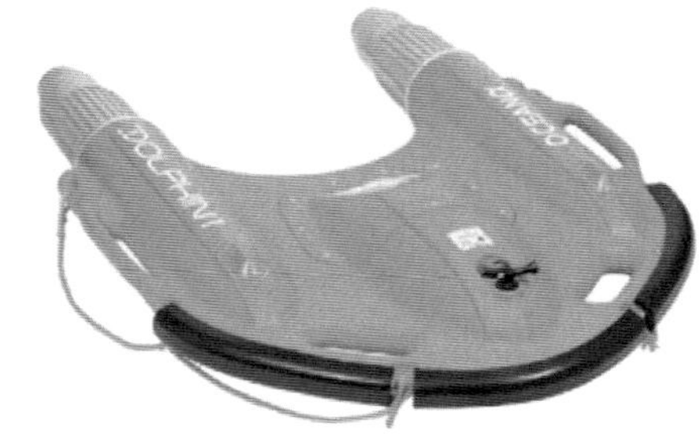

Figure 2: Ocean Alpha's Dolphin1.

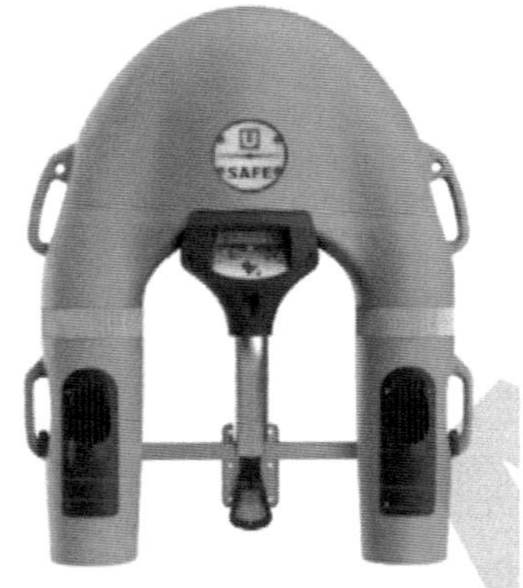

Figure 3: U SAFE Oceania's device.

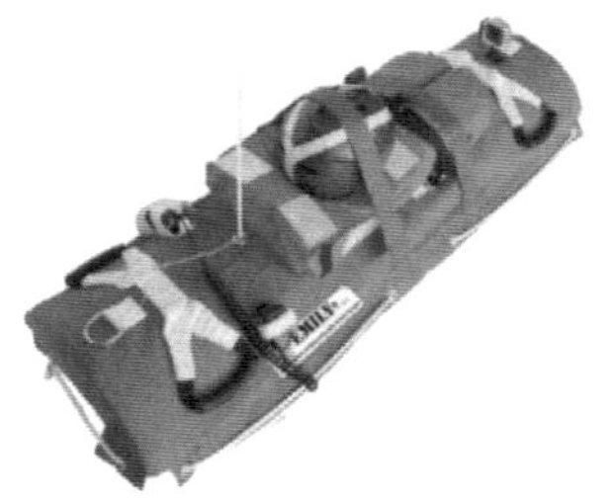

Figure 4: Hydronalix's EMILY device.

2.1 (b) Remotely Piloted Aerial Systems (RPAS)

Remotely Piloted Aerial Systems (RPAS, often referred to as drones). The current focus is on use of in Visual Line of Sight (VLOS) applications.

The Canadian Coast Guard drone capability is still in its early stages. Building on previous testing in 2016, the experimentation effort is intended to support the development of concepts, plans and procedures, and determine how best to deploy such devices to support the Coast Guard mandate and programs.

Testing and evaluation are commencing in early 2021, with reports on the test results due in the autumn. Test events are planned from shore (tethered and untethered), from large vessels and if time and safety allow, from small vessels or craft. A focus on communications and procedures will define some of the constraints of this technology, such as use in limited visibility, weather, or type of target. The appropriate levels of automation and artificial intelligence will be tested on SAR RPAS for search pattern, launch and recovery, and target recognition (where available).

Efforts are ongoing to acquire five (5) small RPAS (under 10kg) supporting VLOS flight operations for the Canadian Coast Guard, with VTOL, capable of operating in relatively harsh conditions (e.g., winds up to 50 km/h. temp to -20C). These drones would have EO / IR (electro-optical / infrared) capability to support SAR personnel during a search. The intent is to deploy such drones from offshore vessels, either as tethered or untethered units.

An additional 20 micro (under 2kg) RPAS are being acquired to further analyse the potential to support personnel and decision-making during response to incidents, including SAR, oil spills and assessment of abandoned or wrecked vessels. These drones are more suited to shore-based launch and recovery, capable of supporting search efforts as an additional set of eyes, again with IR capacity for poor visibility. It is expected that the current state of micro RPAS technology will offer sufficient automation, detect and avoid, GPS location, range and flight time to make drones a solid enhancement allowing the SAR program to save lives more efficiently.

Figure 5: Small RPAS (Drone).

Figure 6: Viewing Coast Guard ship from the RPAS.

Figure 7: Coast Guard large RPAS (Schiebel Camcopter) trials 2016.

2.2 GERMANY – DGzRS – ONGOING R&D PROJECTS BY THE GERMAN MARITIME SAR SERVICES

Thomas Luebcke, DGzRS Head of Research & Development

2.2 (a) ARTUS (2019-2022)

The aim of the ARTUS project is to develop an assistance system for maritime SAR forces that is capable of automatically transcribing VHF radio transmissions and assigning them to potential senders (ship radio stations). This is intended to compensate for both acoustic and cognitive factors that influence communication. At the same time, the system to be developed should make the documentation and traceability of measures taken intersubjectively comprehensible. * The research project will specifically program a deep neural network for the recognition and transcription of radio communication using various data sources. ** In order to be able to test a practical demonstrator at the end of the project, the developed technology must be completely usable offline. For the application in the German Exclusive Economic Zone in the North Sea and Baltic Sea, it must be able to transcribe radio messages in German as well as English. In addition, the project aims to train a deep neural network incrementally for the first time. In this way, it is to be tested whether a system can be enabled to learn during operation and to successively improve its interpretation performance by means of learning loops. In the course of the project, the maritime speech recognition system to be developed will be combined with locating algorithms that have also been newly developed and allow the potential senders (ship radio stations) to be assigned to the transcribed radio messages. For this purpose, the new algorithm will access various nautical data that can be generated on board ships and use these, extended by radio direction finding and voice data, to identify the sending vehicle.

In addition to its use on board ships, the envisaged technology can also be used effectively at the shore-based communications interface or in simulator training.

** This facilitates the keeping of the mission diary by automatically storing all information received via VHF marine radio in exactly the same way as it was received. Possible misunderstandings can thus be prevented and any personal interpretations of the receiver - because this also plays a role in stressful operational situations - can be worked through in a well-founded manner with the help of the ARTUS system recordings in the subsequent debriefing.*

*** Neural networks are a series of algorithms that mimic the operations of a human brain to recognize relationships between vast amounts of data - (Investopedia). A deep neural network is a multi-layer neural network.*

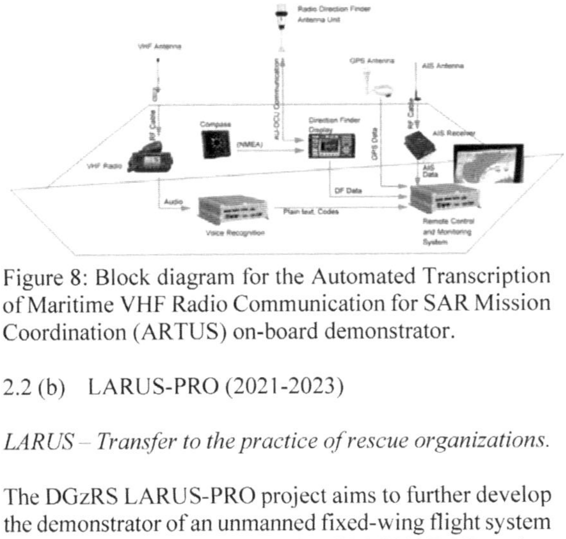

Figure 8: Block diagram for the Automated Transcription of Maritime VHF Radio Communication for SAR Mission Coordination (ARTUS) on-board demonstrator.

2.2 (b) LARUS-PRO (2021-2023)

LARUS – Transfer to the practice of rescue organizations.

The DGzRS LARUS-PRO project aims to further develop the demonstrator of an unmanned fixed-wing flight system developed in the recently completed LARUS R&D project into an operational demonstrator for maritime search and rescue services and to test it under operational conditions over a period of one year. The DGzRS expects this intensive test phase to provide further experience on the benefits, operational tactics and limitations of this potential assistance system. The project will build on the experience gained in the LARUS project in the operation of the flight system. Any need for adaptation identified will be addressed, for example, in the form of a modular payload concept. In addition, a BOS-compliant fixed ground control station is to be developed, the range of the flight system is to be extended by adding satellite communication to the multi-link approach developed in the LARUS project, and the formally required approvals for mission-driven operation of the mission demonstrator are to be obtained. During the trial operation, comprehensive data will also be generated, on the basis of which AI- and ML-based (Artificial Intelligence and Machine Learning based) image recognizers can be trained in the future. The project also intends to initiate citizen-oriented scientific communication on the topic of unmanned aerial systems in civil security in order to raise awareness of the benefits of such innovative systems among a broad section of society.

Figure 9: German rescue drone 'LARUS'.

2.2 (c) Digital Twins for smart SAR units and fleets

With the commissioning of the 28-meter rescue cruiser class of the DGzRS, these ships were consistently equipped with telematics, which makes it possible to transmit more than 500 sensor data from all areas of a ship in real time to shore, store them there and keep them ready for analysis. In the meantime, this has resulted in a considerable stock of mission-driven ship operation data sets, which are now to be used to make ship operation safer and easier to plan in the future. This is because mission operations and the diverse coordination possibilities between machinery, navigation and dynamic environmental conditions are a challenging field of research and application for digital twins. A Digital Twin is defined here as a lifecycle enriched model of a real system and is used to analyze and predict system behavior. Digital twins in this case also allow to combine the very heterogeneous (sensor) information (navigation, weather, cameras, engine, ...) into a unified, coordinated prognostic model. For example, a digital twin for a ship allows early detection of anomalies such as engine problems by comparing expected behavior (forecast of the digital twin) with current sensor information. Prototypes for IT-based ships of a new generation are thus created here. Another focus is the creation of such digital twins for fleets of ships. The aim is to optimize fleet behavior and to investigate the transferability of learned insights from one ship to another. In this project, the focus is on upgrading the ships as IT platforms, creating Digital Twins of the ships and, most importantly, Digital Twins of entire fleets. To verify the methods, various AI services such as anomaly detection and fleet optimization will be implemented. Using real-time data and intelligent forecasts, bunker costs can thus be reduced at the fleet level, operating materials can be procured according to demand, and warehousing can be reduced. Furthermore, the realization of predictive maintenance as well as real-time feedback for the optimization of machine settings is targeted.

2.2 (d) TRAGVIS

Portable range-gated viewing system for vision enhancement in maritime environments.

Searching for people in the water is a central element of many SAR missions. Visibility conditions primarily define how the available rescue units are integrated. For example, especially when visibility is greatly reduced - in fog or heavy snowfall - minimum distances between the individual search units must be maintained for safety reasons (e.g., when searching by formation). Nevertheless, these safety distances often exceed visibility, so that there is a risk of not recognizing persons in the water. The R&D project TRAGVIS is developing a range-gated system especially for the application on small SAR units, which allows to selectively illuminate range windows and to blank out reflecting obstacles (water drops). For this purpose, a pulsed laser-based illumination source is coupled with the shutter of a high-performance camera system. In this way, previously defined depth planes can be captured and made visible. The project is primarily aimed at sensor miniaturization so that range-gated viewing systems can also be used flexibly on smaller search and rescue units in the future.

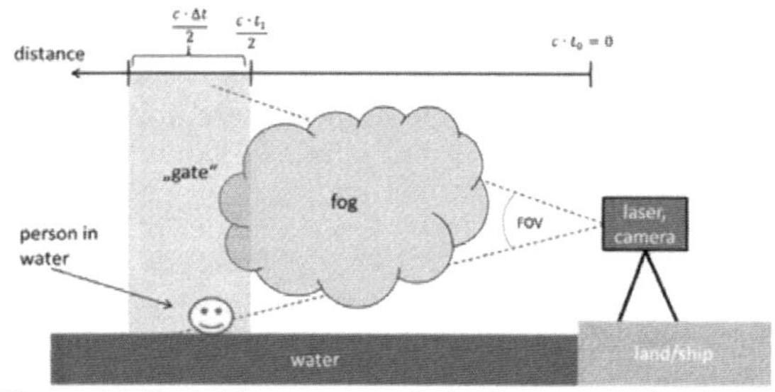

Figure 10: Portable range-gated viewing system for vision enhancement in maritime environments.

2.3 NORWAY – NORWEIGIAN SOCIETY FOR SEA RESCUE – NSSR – THE DIGITAL LIFEBOAT

Martin Fuhr Bolstad, NSSR Project Director

The Norwegian Society for Sea Rescue's vision is that no one should drown and our work to save lives and prevent serious incidents is constantly being developed to achieve this vision. The development and use of digital aids at sea is growing rapidly and especially in the leisure fleet and small fishing boats. Boats are equipped with digital equipment and boaters bring smartphones, smart watches, and other digital equipment. The Digital Lifeboat is about utilizing this equipment to save lives and to prevent accidents. In rescue operations, precise information is of critical importance. Traditionally, emergency alerts are given orally on radio or by mobile phone. It is often difficult to obtain precise information in a stressful situation and it can be difficult to locate the accident precisely enough. Fatal accidents are often associated with a lack of notification or lack of information about where the accident takes place. The Digital Lifeboat is about people being able to use the equipment they have and the solutions they use to warn of distress or that digital solutions themselves must perceive and warn of dangerous situations. The preventive part of the Digital Lifeboat is about being able to reach people where they are with precise information that is relevant to them where they are and in the situation they are in. This can be, for example, danger warnings in changing weather, but it can also be competence-enhancing information about safe traffic at sea. For several years, the rescue company has built up a digital security solution around the RS SafeTrx app. This is built in collaboration with many other sea rescue services in Europe and the rest of the world. With 'The Digital Lifeboat', we will open up this solution so that other digital solutions used at sea can deliver alerts about emergencies and the need for assistance to the rescue service. This measure is about expanding the technological solutions and integrating solutions used at sea in this open network so that users can both notify and receive messages and information that are accident prevention. It also wants to create opportunities for contact between those who travel at sea and to give people the opportunity to help each other with more trivial problems. The measure itself is about establishing the central systems for receiving and handling alerts and sending out information. The handling of alerts also includes being able to distinguish false alerts and to risk assess the alerts based on content and conditions related to the situation (type of vessel, position, weather forecast, etc.) and to be responsible for local spread of relevant information to resources close to the incident and to emergency contacts (set up by uses, such as parents when children are at sea).

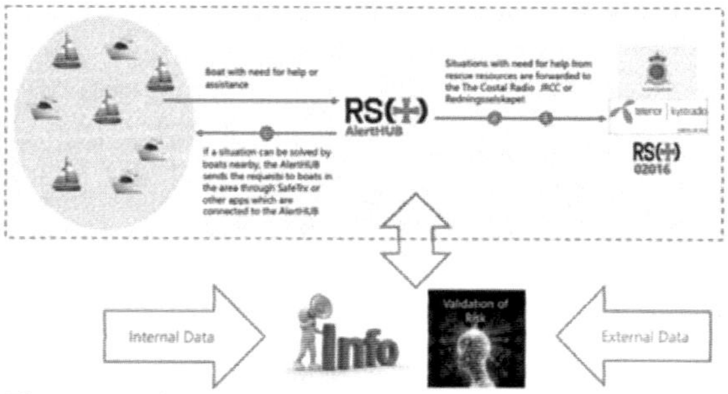

Figure 11: The Rescue Hub system.

The measure will over time ensure that better and more accurate information is collected in connection with accidents at sea. One of the purposes of the measure is to establish solutions for sharing this information with relevant actors within the rescue service to increase the quality of rescue operations.

2.4 SOUTH AFRICA – NATIONAL SEA RESCUE INSTITUTE – NSRI – 'FUTURES COMMITTEE'

Andrew de Kock, Volunteer, NSRI Futures Committee

The NSRI 'Futures Committee', as it has become known, was instigated by and is made up of volunteers from around the country, all having other roles within the Institute. The mandate is to investigate developments and potential disruptors likely to impact the operations of the Institute.

The following are some of the current projects the group is undertaking, with the exception of the COVID 19 social media-based training and JetRIB projects which were instigated and completed by our Operations Department.

2.4 (a) Remote Piloted Aerial Systems (RPAS) – SAR Ops & Beach

The use of drone (RPAS) platforms has grown rapidly in the past few years. This project is investigating the potential for operationalising the use of RPAS within the NSRI including operational, organisational and (especially) regulatory requirements.
The regulations around RPAS operations are in South Africa are very strict and the Institute is required to obtain a commercial RPAS Operator's Certificate to utilise these systems legally. Notwithstanding this, many SAR operations would require waivers to existing regulations.

A key learning to date, apart from in beach/surf patrol, is that many of the operational capability requirements of SAR operations – especially vessel borne operations – are challenging for many RPAS platforms. In addition, there appears to be little or no experience in SAR operations outside fair weather monitoring deployments. This project is investigating various use cases and operational conditions.

Another focus area is around RPAS payload which is where the focus of the platform's utility should be (deploying lifesaving appliances, imagery & image processing & identification, etc.)

Figure 12: Deploying a life-ring from a RPAS (drone).

Figure 13: RCRD operating in heavy waves.

2.4 (b) Remote-controlled Rescue Device (RCRD)

Desk research into this was completed just prior to COVID lockdown in South Africa. The potential use cases focused on swift-water and beach deployments to access swimmers in distress quickly and deployment into locations that were inaccessible to rescue craft or dangerous for rescue swimmers.

Line of sight visibility with the device is critical and given the NSRI operates relatively small craft (up to 15m) this is frequently not possible severely limiting the utility of these devices in our operations.

2.4 (c) "JetRIB" development

Current inflatable and RIB surf rescue craft are almost impossible to operate safely in the surf zone with only the helmsman aboard. Jetski's/PWC's (personal watercraft) meet that requirement but have limited passenger/casualty carrying capacity. The hybrid "JetRIB" allows for two crew to be dropped off with the casualty, then picked up along with the casualty and getting them to safety, even in challenging surf conditions.

Another key advantage is the absence of an exposed propeller, so the injury risk is removed, and performance degradation challenges of prop-guards is no longer a factor.

Figure 14: NSRI 'JetRIB'

2.4 (d) Vessel telematics

This is a very broad project with key streams focusing on connectivity (2-way data transmission, data collection, etc.), operations (command & control, navigation, search pattern & engine monitoring, etc.); asset management (maintenance, fuel usage, engine hours, etc.) and operational monitoring and strategic planning (location, duration, nature of operations, average cost, etc.)

More and more devices on our craft are able to communicate data, but a data-driven organisation needs to exploit that data to gain any value from it. 2-way data transmission, which facilitates increased shore support for craft at sea, reducing crew workload and increasing focus on their primary activity (seeing and rescuing casualties), is one bottleneck that needed to be resolved. Satellite-based options exist but are expensive, VHF solutions have coverage limitations given our geography, and newer technologies like Mesh still need investigation. Given that the majority of our operations are not far offshore, an immediate, affordable solution was installing on-board cell-based Wi-Fi routers.

Apart from data-transmission there were a number of immediate and unexpected benefits, such as simplified 'SITREP' updates via WhatsApp allowing image, text and voice capabilities even in some VHF 'blackspot' areas.

The goal of this project is to investigate various solution possibilities which cover the above, but importantly also aggregate the data into the Institute's platform in order store appropriate data centrally and to facilitate maximised utilisation of the data.

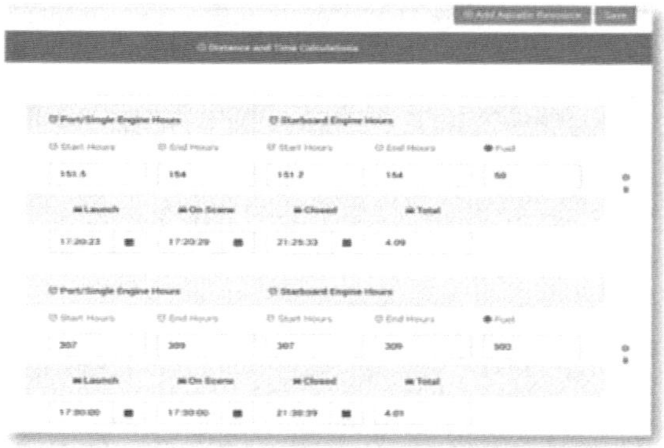

Figure 15: Example of Telematics Data output part 1.

Figure 16: Example of Telematics Data output part 2.

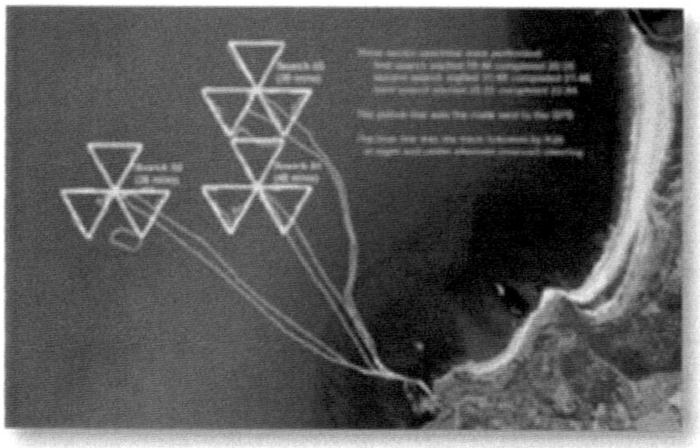

Figure 17: Example of Telematics Data output part 3.

2.4 (e) Training & community of volunteer crew under strict COVID-19 lockdown

The strict and prolonged COVID-19 lockdown in South Africa prevented any practical training or physical contact between the crew and brought the Training Department's training program to a grinding halt.

Within days, the team had envisioned a range of virtual training activities which they implemented almost immediately using social media, setting up a private Facebook group. Other technologies were available to deliver the activities (e.g., MS Teams) but the team recognized that using mainstream social media would remove barriers to entry.

In addition to specific training series on various topics, 'Rescue Radio' interviews with various crew and personalities in the NSRI have allowed many stories to be told, recorded and thereby documented a part of the history that would likely never have been known.

Since the lifting of lockdown restrictions, these activities have continued though at a reduced tempo, but are still very popular with the crew.

2.5 SWEDEN – SWEDISH SEA RESCUE SOCIETY – SSRS – HOW MIGHT WE DEVELOP AN EMISSION-FREE RESCUE BOAT?

Fredrik Falkman, SSRS Head of Innovation

Due to a grant from the Postcode Lottery's Dream Project fund, SSRS has received funding to carry out an environmental project to investigate how the organization can conduct life-saving activities more environmentally friendly and with greatly reduced emissions. The project is planned to run between 2020 and 2023.

SSRS has historically developed several boat types in different sizes which are used successfully in the business. Common is that everyone uses internal combustion engines adapted for petrol, diesel or a diesel-like fuel. Previous development of hulls and propulsors has therefore to a large extent been able to rely on fuel tanks in which it is possible to carry a lot of energy. The dream project's starting point is to greatly reduce emissions by trying new technologies beyond internal combustion engines. One of the project's main goals is to develop a lifeboat prototype that carries out the assignments, which instead places a higher requirement on energy efficiency, therefore both a new hull and a new type of propulsor needs to be evaluated. SSRS Annual Report 2020 expresses the most critical point to bridge with regard to the lifeboat prototype as follows:

"If a fuel or an engine does not meet the technical requirements to be able to quickly rescue people in need, it is something the Sea Rescue Society cannot consider as an alternative."

The overall purpose of this project is to investigate how SSRS emissions, mainly greenhouse gases, can be greatly reduced. This includes developing, building and testing a lifeboat prototype that gives rise to a lower level of emissions while reducing the need for maintenance. It also includes a survey of the emissions that the organization currently gives rise to and a plan how to reduce these. The results will then be presented where a central part is the presentation of a lifeboat prototype. The presentation will also spread knowledge and experience outside SSRS to help others, e.g., the leisure boat industry and SSRS sister organizations, move towards a more environmentally friendly development.

The development steps in the Emission-Free Rescue Boat project include:

- Initially looked to Hydrogen fuel-cells. Preliminary conclusion: 5-10 years too early – wait for the trucking sector to industrialize.
- Diesel ≈ 50x energy/weight compared to current batteries - careful study of mission profile + optimizations necessary.
- Thorough analysis of historic mission reports:
    a. Rescue Boats < 8 meters ≈ 50% of missions
    b. > 95% of those fit into this preliminary mission profile
    c. 15 Nm dash, 15 Nm light tow, 10 Nm "quick" return, 20 Nm limp home
- Also developed cloud-connected data-logging solution to get more granular data for future fine tuning of mission profile. Installation into Rescue Fleet commenced.
- Currently designing 4 baseline hull concepts to inform further work:
    a. Conventional V-hull
    b. Semi-planing
    c. Inherently stable retractable Hydrofoil
    d. Inherently unstable retractable Hydrofoil
- Expecting to commence prototype building late 2021.

As an interim step, in 2019 the SSRS commissioned the 14-metre (46-ft) composite rescue vessel 'Rescue Mercedes Eliasson Sanne'. She carries a crew of three or four people for rescue missions and operates out of the Rörö rescue station in Gothenburg's northern archipelago. She is the first-ever in the Sea Rescue Society's fleet to have parallel diesel and electric motors. The primary powertrain is twin 650 horsepower Scania diesels with water jets. The electric drive system consists of dual Torqeedo Deep Blue 50KW electric motors for a combined 160 horsepower-equivalent with two 10 kWh Torqeedo Deep Blue lithium batteries. The vessel can operate silently and emission-free up to 1.5 hours between recharges at a search speed of 6 knots.

"We use the electric mode when entering and leaving port through Sweden's environmentally sensitive waterways," said Lars Samuelsson, head of the Maritime Department. "The crew uses the diesels for high-speed runs out to the rescue location, then switches back to electric for search and rescue operations. It is crucially important to go quietly on the job since even the smallest sound can hinder the search for a missing person in the water." [7]

Figure 18: 'Rescue Mercedes Eliasson Sanne', Torqeedo electric outboards in housings P&S aft.

2.6 UK/REPUBLIC OF IRELAND – ROYAL NATIONAL LIFEBOAT INSTITUTION – RNLI – HUMAN FACTORS IN THE DESIGN OF LIFEBOATS

Tom Coe, Whole Body Vibration, RNLI Noise and Ergonomics Project Manager

Building on the historical work it has contributed towards in this field, in recent years, the RNLI has invested in bringing human factors expertise in-house to enhance the safety and useability of their lifeboat fleet [8]. Human factors (HF) is the scientific discipline and profession focused on optimising system performance and human well-being through design. While vital to producing successful outcomes within all design work, consideration of the needs of users is especially critical in high hazard industries, such as marine search and rescue. The core focus of human factors work at the RNLI is on protecting the crews who selflessly volunteer to go to sea on behalf of the charity, risking their safety to help those in need. This is in addition to ensuring those who manufacture and maintain the boats can do so safely.

The degree to which human factors work impacts a design can vary greatly and input has been found to be of value throughout the design life of each vessel. There are many tools used to make these decisions such as physical assessment methods, anthropometric data, and user fitting trials.

As well as applying these industry-standard techniques, the RNLI has also developed their own in-house digital human modelling capabilities. This human model bridges the gap between un-poseable digital human models, such as those presented by Dobbins et al [9] and more advanced commercial solutions such as JACK™, developed by Siemens.

Figure 19: Measuring for the human models.

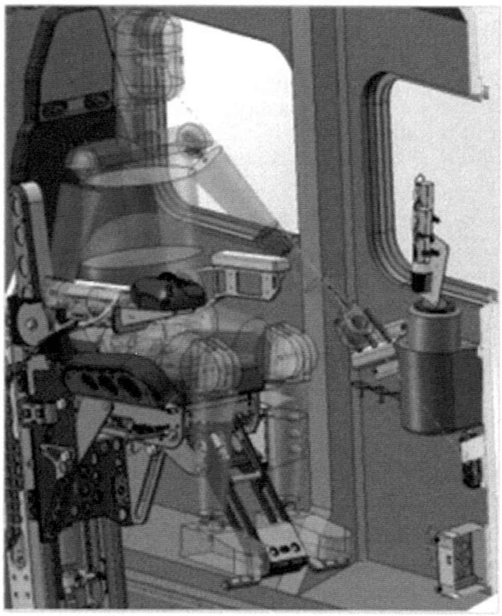

Figure 20: Digital human model; locating engine controls.

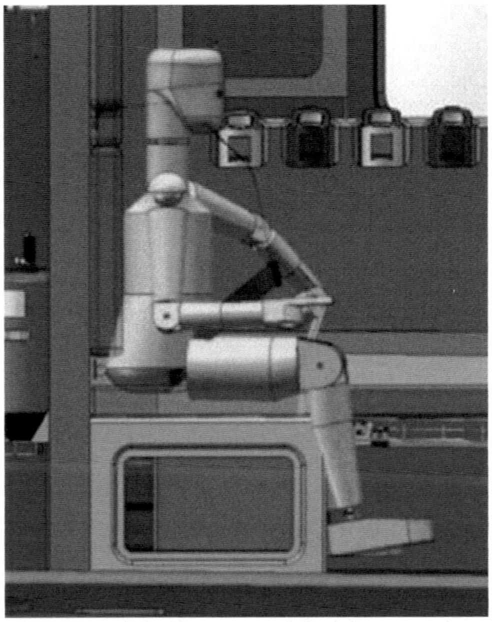

Figure 21: Digital human model; sizing a seat harness.

The combination of digital human modelling and other technical methods have been used to assess human factors interventions in both in-service vessels, making small changes to improve crew-machine interface on Rigid Inflatable Boats, and in the development of the wheelhouse design for the life extended Severn Class All Weather Lifeboats.

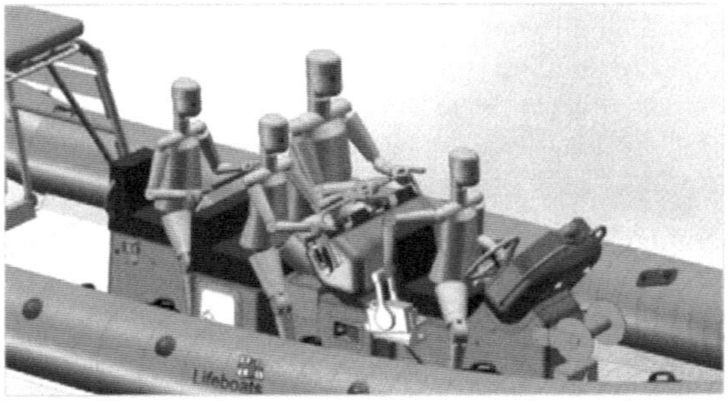

Figure 22: Digital human models; Seat and controls layout on RHIB.

Historically at the RNLI, human factors work has been utilised by the personal judgement of engineers and a basic toolset, with support from external consultants. The decision to further invest by incorporating human factors engineering ensures the ongoing safety and improvement of RNLI vessels, and crewing can increasingly become more inclusive.

2.7 AUSTRALIA – YELLOW SCORPION PTY. LTD.

Jim Bethell, Search and Rescue Consultant, Yellow Scorpion Pty. Ltd. Perth, WA.

Electro Optic Searching

Within maritime Search and Rescue (SAR) operations there is a big swing towards Electro Optics (EO) to assist us with our searching. By day, that is as simple as a camera on an aircraft or drone, by night we use Night Vision Goggles (NVG) and Infrared (IR) sensors. These systems are either used by an observer or are increasingly being used by image recognition software to detect and identify targets.

Visual maritime searching is a well-studied and documented process that gives us planning tools and a probability of detection for any given scenario (as we find in the IMO IAMSAR Manual). We do not have anything similar for electro optic searching which makes it impossible to effectively task search assets or give reliable data regarding the effectiveness of our search.

Yellow Scorpion has been working with SAR tasking authorities and operators to improve EO search techniques for any given sensor and is working towards a model that can be used in the same way we model visual searches. We found that most sensors are looking beyond the maximum visual detection range for the target, which is pointless.

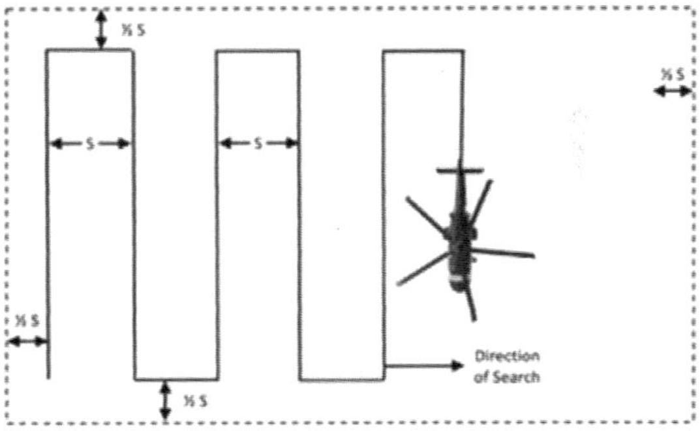

Figure 23: Recommended search pattern.

With some simple calculations and some computer modelling we have been able to visually represent the maximum object Detection, Recognition, and Identification (DRI) ranges in order to ensure that the sensor is being used as efficiently as possible and giving an estimate of search sweep widths. This allows operators to select their optimum sensor for the conditions and create and effective search and allows us to model how long a target is in view and detectable to the sensor for any given variable.

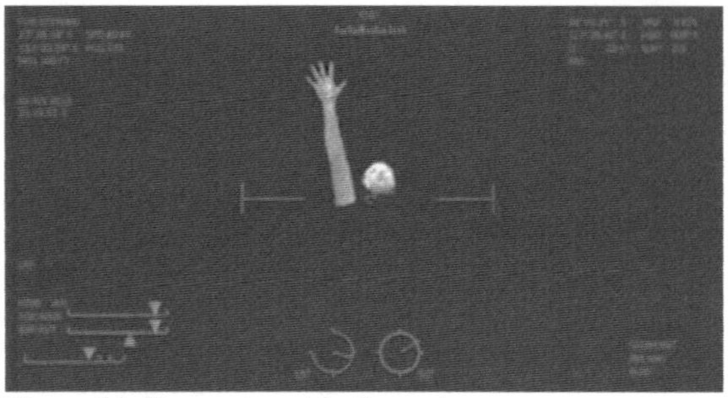

Figure 24: Human target in the water.

2.8 REPUBLIC OF IRELAND – SAFETRX

Nessa Malone, SafeTrx Product Manager, 8 West Consulting Ltd., Cork

SafeTrx – Taking the 'Search' out of Search & Rescue.

Designed in close co-operation with Police, Coast Guard and Marine Rescue (SAR) professionals from across the globe, SafeTrx has become the leading maritime non-GMDSS safety system. It gives national SAR authorities a simple means of tracking and assisting inshore boat users using Smartphone and wearable apps.

Figure 25: RYA SafeTrx emergency alerting system.

SafeTrx is used by Police, Coast Guard and Marine Rescue authorities in Australia, the Netherlands, Germany, Spain, Norway, South Africa, Finland, the UK, Ireland, Dubai, and Denmark. SafeTrx has been involved in 1,000 rescue and assistance missions in 2020, and its usage is steadily increasing. The following organisations use the SafeTrx system to safely monitor and respond to leisure sailors assistance and distress calls:

- Koninklijke Nederlandse Redding Maatschappij (Royal Netherlands Sea Rescue Institution)
- Deutsche Gesellschaft zur Rettung Schiffbrüchiger (German Maritime Search & Rescue Service)
- Redningsselskapet (Norwegian Sea Rescue Society)
- Suomen Meripelastusseura (Finnish Lifeboat Institute)
- Sociedad de Salvamento y Seguridad Marítima (Spanish Maritime Safety Agency)
- Sofartsstyrelsen (Danish Maritime Authority)
- National Sea Rescue Institute (South Africa)
- Dubai Police
- Australian Volunteer Coast Guard
- Marine Rescue Queensland
- Royal Yachting Association
- Irish Sailing Association
- Sjöräddningssällskapet (Swedish Sea Rescue Society) *
  *Implementation underway

Each year, the SafeTrx team host the SafeTrx User Group meeting where they meet with Coast Guard, SAR organisations and boating associations to help shape the technical and operational aspects of the platform. Working together, their objective is to continue to evolve SafeTrx.

The SafeTrx system is open to accommodating high quality solutions into the ecosystem and integrates with other systems to help expedite search and rescue efforts.

- The SafeTrx platform already receives alerts from third party IoT devices and apps such as RS Aktiv Nødknapp, Seaber mobile app and Fell Marine MOB+ device.
- SafeTrx is integrated into the SafeTrx Incident Management System (IMS) which is used for tasking lifeboat crews to respond to emergencies. The one-way integration with SafeTrx allows trips and emergency call details to automatically create an Incident within IMS.
- SafeTrx also has the ability to send trip and emergency call data to SafeTrx Flotilla Event, which gives race organisers the ability to plan, manage and deliver live tracking on a web-based race viewer. Race participants real time locations are available to race organisers and the Search and Rescue (SAR) authorities in case of emergencies.

SafeTrx app on the mSafety wearable from Sony.

Figure 26: mSafety wearable emergency alerting device.

Innovation is central to the ongoing development of SafeTrx. For the last 18 months, SafeTrx has been working as a development partner with Sony Network Communications to implement SafeTrx on the new mSafety wearable platform from Sony. mSafety is an Internet of Things wearable operating on the LTE-M cellular network.

There are some unique aspects associated with this wearable compared to other technologies that we have investigated:

1. Dedicated wearable (no 3rd part apps that can consume battery and data)
2. Encrypted end to end communications
3. Long-range using LTE-M cellular technology
4. Designed as a low power device

The SafeTrx team has built a completely new application on this new Sony architecture with the following capabilities:

- Distress and Alert Transmission: The SafeTrx app on the wearable is integrated directly with the SafeTrx Command and Control (C2) system, meaning tracking and alerting data gets sent directly to the designated monitoring authority.
- Always Connected: The mSafety wearable comes equipped with a built-in eSIM allowing it to roam onto any LTE-M telecommunications network, thereby providing the SafeTrx app with seamless connectivity. Unlike other smartwatches, there's no need to carry a phone to transmit location data from the wearable.
- Longer Range: SafeTrx on mSafety from Sony transmits at ranges far in excess of smartphones or other non-IoT wearables. Independent testing by the Dutch telecommunications company KPN show the device continuing to transmit 17km from the nearest cell tower, at –134db RSRP.
    - Lightweight and Water Resistant: The Sony mSafety wearable weighs just 40g, light in comparison to other devices. It is IP68 certified which means that it is resistant to continuous submersion up to a maximum depth of 1.5m underwater, for up to thirty minutes.
    - Ease of Use: The SafeTrx app has a simple User Interface that individuals in stressful circumstances can easily understand.

The SafeTrx app on the mSafety wearable from Sony is currently in beta trials with SAR organisations in the Netherlands, Norway, and Sweden. Trials are particularly focused on how SafeTrx can improve crew member and lone worker safety.

3. CONCLUSION

Many of the maritime search and rescue organizations can trace their origins back to a time when the loss of life at sea was far too common. Since time immemorial humans have had beneficial interactions with the sea – for food, transport, and pleasure. However, these interactions have not always been without cost – the risk of death has always been present. In the 1820s there were about 1800 shipwrecks each year around the British Isles alone. Amongst maritime folk the risks of going to sea were understood; significant loss of life was expected. Often in sight of a coastal community, a sailing ship would be blown aground, the crew would take to the rigging to try to survive, townsfolk would see the plight of the hapless sailors and often would put to sea, in whatever craft they may have at hand, at significant risk to their own lives. These occurrences eventually led to the first purpose-designed lifeboat – the 'Original', built on the North-East coast of England in 1789. Similar tragedies led Sir William Hillary to call, three decades later, for the establishment of a national organization to perform rescue around the coasts of the British Isles. The Royal National Lifeboat Institution (RNLI) was followed during the 19th century by similar organizations in many other countries – many private charities like the RNLI, some government funded organizations [3].

Over the past century improvements in technology (as well as safety standards, enforcement of regulations and education) have dramatically reduced commercial ship casualties. The risks of capsizing, foundering, grounding, collision, and fire, have all been reduced for commercial vessels.

During the past century dedicated rescue resources have also improved dramatically. At the end of the 19th century, rescue lifeboats were open boats propelled by sail and oars. Modern rescue lifeboats, and larger sea rescue cruisers, have speeds of 25 knots or more, extended range, modern electronics for searching, navigation and communication, excellent sea-keeping for operation in heavy weather, and can operate in conjunction with other rescue resources, such as aircraft and helicopters.

See Figures 27 to 30 are examples of rescue craft, then and now.

THEN

Figure 27: A RNLI nineteenth century 'pulling' lifeboat (RNLI website).

Figure 28: The first RNLI steam lifeboat 'Duke of Northumberland',1890 (RNLI website).

AND NOW

Figure 29: Modern Norwegian high speed rescue vessel (LK Enger photo, NSSR).

Figure 30: Modern 28 M German high speed sea rescue cruiser with daughter boat (Fassmer photo; DGzRS 28 M).

Aircraft can fly in extreme weather conditions, travel great distances to provide oversight, and possibly deploy liferafts. Helicopters can provide rescue over reasonable distances (potentially being refueled in flight), and in areas where it may be risky to attempt a rescue by ship, such as near shore. Search and Rescue (SAR) Technicians, an elite group of primary care paramedics, can provide on-scene medical aid and evacuation: they can parachute from aircraft or be lowered from helicopters to assist survivors until recovery is possible.

The IMRF Future Technology Panel works to build on the progress made in Search and Rescue capabilities during recent decades, promoting the sharing of the development and implementation of technologies which can enhance Search and Rescue capabilities. We have outlined in this Paper some of the wide range of technologies being investigated and shared by the member organizations of the Future Technology Panel, successful implementation of these technologies will improve the safety of all persons on the sea.

However, technology alone is not a cure all. Finally, and perhaps most importantly, are the competent and capable personnel who have the courage and dedication to be willing to assist their fellow human beings:

"...The loss of the Solomon Browne and her crew was caused in the aforesaid circumstances in consequence of the persistent and heroic endeavours by the coxswain and his crew to save the lives of all from the Union Star. Such heroism enhances the highest traditions of the Royal National Lifeboat Institution in whose service they gave their lives." [10]

Hopefully, Future Technologies will also make the lives of rescue personnel safer.

4. ACKNOWLEDGEMENTS

I acknowledge the encouragement and contributions from many of the world's Search and Rescue Organizations. I also acknowledge the support of my Dalhousie University colleagues, and the assistance from Tara Sherrard, B.Eng., former Industrial Engineering student, in preparing this Paper.

I particularly acknowledge the inspiration from the many front-line Search and Rescue personnel, whose dedication and self-less efforts remind me that there are so many good people in this world.

5. CONTACTS

5.1 INTERNATIONAL MARITIME RESCUE FEDERATION

https://www.international-maritime-rescue.org/
Email: info@imrf.org.uk
Phone: +44 (0) 1569 767405
Mail: PO Box 1389
Enfield, EN1 9GF
United Kingdom

5.2 EDITOR

John W. Dalziel, P.Eng., MRINA,
Adjunct Professor, Industrial Engineering,
Dalhousie University, Halifax, Nova Scotia, Canada - john.dalziel@dal.ca

5.3 CONTRIBUTER'S WEBSITES

5.3(a) SAR Organizations

- Canada – Canadian Coast Guard – CCG https://www.ccg-gcc.gc.ca/index-eng.html
- Germany – German Maritime SAR Service – DGzRS

https://www.seenotretter.de/en/who-we-are/what-we-do/
- Norway – Norwegian Society for Sea Rescue – NSSR
https://www.redningsselskapet.no/english/
- South Africa – National Sea Rescue Institute – NSRI
https://www.nsri.org.za/
- Sweden – Swedish Sea Rescue Society - SSRS
- https://www.sjoraddning.se/information-english
- http://www.surtsey.org/
- UK / Republic of Ireland – Royal National Lifeboat Institution - RNLI
https://rnli.org/

5.3(b) Companies

- Australia – Yellow Scorpion Pty. Ltd.
https://www.yellowscorpionaviation.com
- Republic of Ireland – SafeTrx; 8 West Consulting Ltd.
https://www.safetrxapp.com/products/safetrx-maritime

6. REFERENCES

[1] BUTLER, G., 'The Heroes of Isle aux Morts' [image], Walsh, A., author, Tundra Books, February 2001.
[2] CANADIAN COAST GUARD, 'CCCG Fleet: Vessel Details – CCGS Ann Harvey', Government of Canada Website, 2021.
[3] ROYAL NATIONAL LIFEBOAT INSTITUTION, '1824: Our foundation', Royal National Lifeboat Institution Website, 2021.
[4] INTERNATIONAL MARITIME RESCUE FEDERATION, 'The IMRF & Our Mission', International Maritime Rescue Federation Website, February 2019.
[5] INTERNATIONAL MARITIME RESCUE FEDERATION, 'Why not join the IMRF Future Technology Panel', International Maritime Rescue Federation Website, October 2020.
[6] PRESENTATIONS Section: Text, diagrams and pictures from the contributing organizations, unless otherwise credited.
[7] SMALLRIDGE, T., 'Swedish Sea Rescue Society boat is first with diesel and Torqeedo electric power', Marine Business World Website, January 2020.
[8] DOBBINS, T., ROWLEY, I. & CAMPBELL, L., 'High Speed Human Factors Engineering Design Guide', ABCD-TR-08-01 v1.0, January 2008.
[9] DOBBINS, T. et al., 'FRC International Human Factors Integration Conference', 2013, Poole, UK.
[10] ROYAL NATIONAL LIFEBOAT INSTITUTION, 'Penlee Formal Investigation', Royal National Lifeboat Institution Website, 1983.

8. AUTHOR'S BIOGRAPHY

John W. Dalziel, M.Sc., P.Eng., MRINA, Adjunct Professor, Department of Industrial Engineering, Dalhousie University, Halifax, Canada. He studied Naval Architecture in the United States and Scotland, and Management Science in England and has sailed through the Arctic with the Canadian Coast Guard, and with the fishing industry. Over the past half century, most of his experience has been in the supervision of ship construction, refits, and repairs, and in safety inspection and survey on behalf of Flag State and Class. He has spoken internationally on maritime safety, including at the World Maritime Rescue Congresses in Vancouver, Canada, 2019 and Bremerhaven, Germany, 2015. As an Adjunct Professor in the Department of Industrial Engineering, Dalhousie University, Halifax, Canada, he is investigating 'The Application of Unmanned Marine Vehicles to Maritime Search and Rescue'. For many years he has been a Supporter of the International Maritime Rescue Federation and has been participating in the Future Technology Panel since 2016.

Other Contributors: Are technical personnel with their respective organizations.

The Royal Institution of Naval Architects
8-9 Northumberland Street
London, WC2N 5DA
United Kingdom

CURRAN ASSOCIATES INC.
proceedings.com

ISBN 978-1-7138-8307-4